Introduction to Engineering Design and Problem Solving

McGraw-Hill's *BEST*—Basic Engineering Series and Tools

Introduction to Engineering Design and Problem Solving

M. David Burghardt
Hofstra University

Boston Burr Ridge, IL Dubuque, IA Madison, WI New York San Francisco St. Louis
Bangkok Bogotá Caracas Lisbon London Madrid
Mexico City Milan New Delhi Seoul Singapore Sydney Taipei Toronto

WCB/McGraw-Hill

A Division of The McGraw-Hill Companies

INTRODUCTION TO ENGINEERING DESIGN AND PROBLEM SOLVING

1 2 3 4 5 6 7 8 9 0 DOC/DOC 9 3 2 1 0 9 8

ISBN 0-07-012188-5

Vice president and editorial director: *Kevin T. Kane*
Publisher: *Tom Casson*
Executive editor: *Eric M. Munson*
Developmental editor: *Holly Stark*
Marketing manager: *John T. Wannemacher*
Project manager: *Christine A. Vaughan*
Production supervisor: *Heather D. Burbridge*
Freelance design coordinator: *JoAnne Schopler*
Supplement coordinator: *Carol Loreth*
Compositor: *Publication Services, Inc.*
Typeface: *10/12 Century Schoolbook*
Printer: *R. R. Donnelley & Sons Company*

Library of Congress Cataloging-in-Publication Data

Burghardt, M. David
    Introduction to engineering design and problem solving / M. David
Burghardt
      p. cm.
    Includes index.
    ISBN 0-07-012188-5
    1. Engineering design. 2. Engineering mathematics. I. Title.
TA174.B874   1999
620'.0042—dc21   98-18378

http://www.mhhe.com

In loving memory to Henriette "Harry" Vanderryn

About the Author

M. David Burghardt is a professor in the engineering department of Hofstra University. He received his Ph.D. in mechanical engineering from the University of Connecticut in 1971 and is a Professional Engineer in New York state and a Chartered Engineer in the United Kingdom. He is the author of several texts on thermodynamics, diesel engines, and freshman engineering. His current research interests are in two primary areas, lightweight diesel engines and improving technological literacy in elementary and secondary schools. He has coordinated the introduction to engineering experience at Hofstra for the past 15 years and his primary teaching responsibilities are in freshman engineering, thermodynamics and technology, and public policy.

Foreword

Engineering educators have had long-standing debates over the content of introductory freshman engineering courses. Some schools emphasize computer-based instruction, some focus on engineering analysis, some concentrate on graphics and visualization, while others emphasize hands-on design. Two things, however, appear certain: no two schools do exactly the same thing, and at most schools, the introductory engineering courses frequently change from one year to the next. In fact, the introductory engineering courses at many schools have become a smorgasbord of different topics, some classical and others closely tied to computer software applications. Given this diversity in content and purpose, the task of providing appropriate text material becomes problematic, since every instructor requires something different.

McGraw-Hill has responded to this challenge by creating a series of modularized textbooks for the topics covered in most first-year introductory engineering courses. Written by authors who are acknowledged authorities in their respective fields, the individual modules vary in length, in accordance with the time typically devoted to each subject. For example, modules on programming languages are written as introductory-level textbooks, providing material for an entire semester of study, whereas modules that cover shorter topics such as ethics and technical writing provide less material, as appropriate for a few weeks of instruction. Individual instructors can easily combine these modules to conform to their particular courses. Most modules include numerous problems and/or projects, and are suitable for use within an active-learning environment.

The goal of this series is to provide the educational community with text material that is timely, affordable, of high quality, and flexible in how it is used. We ask that you assist us in fulfilling this goal by letting us know how well we are serving your needs. We are particularly interested in knowing what, in your opinion, we have done well, and where we can make improvements or offer new modules.

Byron S. Gottfried
Consulting Editor
University of Pittsburgh

ix

Preface

Introduction to Engineering Design and Problem Solving is designed to help beginning engineering students gain a better perspective on engineering, particularly the creative aspects of engineering design coupled with the rigors of analysis. Engineering design and the design process are generally not well understood, and the text focuses on them through discussions of the design process, examples of student work, and words of wisdom from practicing engineers. Engineering analysis is integral to the design process, and fundamentals in electrical engineering, mechanics, and energy are part of the knowledge base necessary for successful designs.

The first chapter examines the human-made world, the world we live in, created by engineers. The processes employed by engineers to create this world embrace science, the study of the natural world, and mathematics, the study of logical constructs, and include elements uniquely their own. There are philosophical values associated with mathematics, science, and engineering, some similar, others divergent. These values will be explored. One of the purposes of Chapter 1 is to set the stage for the later chapters, providing a context for the designs that we create. For instance, in mathematics and science the concept of uniqueness, a correct answer, is vital. The square root of 25 is 5, not about 5, and similarly, the composition of water is H_2O, not approximately that. However, in the human-made world, optimum or best solutions are important, uniqueness is not. Trade-offs are always made among cost, materials, aesthetics, and other factors. In addition, there may be constraints to be satisfied on the problem in terms of product size, time, and personnel.

Chapter 2 discusses and analyzes the design process. The iterative nature of the design process—problem statement, specifications/clarifications, investigation, brainstorming, creating solutions, evaluation of solutions, and selection of the optimum solution—are examined and illustrated. Creativity is an important aspect to design, and traits and attitudes of creative people are portrayed to see how we can expand our own inherent creativity. The concepts of invention and innovation are introduced along with multicriteria analysis techniques. The area of ecological design is of growing importance nationally and internationally, and the challenges that it brings are examined.

In Chapter 3 design documentation is explored in its several forms. Engineers document their daily work in a design journal which provides a resource for attorneys filing for patents. The presentation of a created artifact is important, hence the design portfolio and design report are discussed. The design portfolio is a guide which assists students in their first designs; it asks what the problem is, what investigations were carried out in implementing the design process. This is illustrated with an actual student portfolio and its assessment. Once a student has worked with the design portfolio, a design report can be written based on the information in the portfolio.

Chapter 4 puts it all together. Engineering analysis in electrical, mechanical, and energy fundamentals is presented along with critiqued design project reports. The area of electrical engineering includes steady-state dc circuits with applications and logic diagrams and circuits. Engineering mechanics includes two-dimensional forces and moments including the use of free-body diagrams. Strength of materials is introduced; with stress and strain the concept of factor of safety is introduced. Energy analysis includes the conservation of mass and energy for open and closed systems. Examples of extended energy analysis are presented. Several fundamental software applications available to engineers, essential for today's engineering practitioner, are examined.

What is it like in the real world of engineering? Chapter 5 addresses this question through extended interviews with six engineers from various fields. These engineers provide insight into how the business of engineering functions, what to be alert to and proficient in. The excitement of engineering comes alive through their stories, why they chose engineering, career moves, what to look out for, abilities to hone.

Much of engineering analysis rests on the foundation of algebraic word problems—how do you set up and solve equations from information presented in words? This is often a challenge for beginning engineering students, and Appendix A reviews basic algebraic and trigonometric word problems with engineering applications. A review of binary number systems is presented.

Many people have contributed to the development of this text, none more important than my students in freshman engineering. I would like to particularly point out the four students who have assisted me by contributing their design projects for all to see—John Buhse, Joseph DiBiasi, Nancy Forsberg, and Ciro Poccia. In addition I appreciate the critical and very valued review of the manuscript by Dr. J. Taylor Beard, University of Virginia; Dr. Diane Beaudoin, Arizona State University; Dr. David L. Clark, California State Polytechnic University; Dr. Rajiv Kapadia, Mankato State University; Dr. Terry L. Kohutek, Texas A&M University; Dr. John A. Krogman, University of Wisconsin-Platteville; Dr. Michael E. Mulvihill, Loyola Marymount University; Dr. Melinda Piket-May, University of Colorado; and Dr. Steven Yurgatis, Clarkson University. The text is very much improved by their individual and collective wisdom.

The overall purpose of the book is to interest and excite students about engineering, particularly with the creative challenge and reward of engineering design. It provides a window on the real world of engineering practice, a world of marketplace demands and tremendous opportunities.

Contents

Chapter 3   Design Documentation   47

Chapter 4   Engineering Analysis and Design   73

1

Understanding the Human-Made World

Engineering and the artifacts of engineering—technology—have been around since the beginnings of human civilization, far longer than mathematics and science; yet most people are terribly ignorant about engineering and technology. In contrast to mathematics and science, engineering and technology are typically not taught in schools in the United States, and the processes creating the enveloping human-made world are largely not understood.

There are many definitions of *engineering,* such as the one from Webster's *College Dictionary:* "the practical application of science and mathematics, as in the design and construction of machines, vehicles, structures, roads and systems." This definition belies the uniqueness of engineering, the body of thought it embodies, and the methodology, uniquely its own, which it employs. Perhaps a more general definition would be a course of study followed by a professional life devoted to the creative solution of problems. Yes, engineers are creative problem solvers, often imagining and designing new technologies as a means to solve problems. Advances that improve the standard of living of people are directly attributable to engineering, from machines that reduce physical labor to systems that provide clean water.

Integration of Mathematics, Science, and Engineering

Table 1.1 indicates some of the differences between mathematics, science, engineering, and the social sciences and humanities. Of course, these are thumbnail sketches, but they will serve to highlight the differences between the disciplines. Science is the study of the natural world, a discipline engaged in discovering the whys and wherefores of natural phenomena. What creates lightning? Why do electrons have a negative charge? There is a process for this investigation—scientific inquiry—in which a hypothesis is posed and logical investigations are undertaken to confirm or deny the hypothesis. Through such investigations, theories are confirmed that explain natural phenomena. Mathematics is used by engineers and scientists as part of their investigations, and it has its own philosophy and patterns. Mathematics is often used to represent a natural phenomenon or human-made device. For instance, Newton's second law of motion—force is the time rate of change of momentum—is often expressed as $\mathbf{F} = m \cdot \mathbf{a}$, a mathematical representation of the law. There are rules of mathematical analysis, called theorems, that allow us to manipulate such equations, such as the commutative law that allows us to represent it as $\mathbf{F} = \mathbf{a} \cdot m$.

The social sciences and humanities provide an entirely different view of the world, a world shaped by human perceptions and understandings. For instance, a novel or a political or social event can be analyzed from many different perspectives. A woman may analyze it differently from a man, an urban dweller differently from a rural person, a minority member differently from a majority member, a technical person differently from a nontechnical person. These are the different lenses through which events are viewed, and opinions shaped. We often realize there is no one correct answer, but the superposition of many beliefs creates the societal belief collage.

What discipline brings all these together? Engineering! In creating the human-made world, knowledge from science, mathematics, and social sciences and humanities must be brought to bear. As contrasted with scientific inquiry and mathematical analysis, engineering design does not seek a unique or correct solution, but rather seeks the best or optimum solution after a variety of factors are weighed, such as cost, materials, aesthetics, and marketability. We will discover that the design process is iterative, creative, nonlinear, and exciting. The solutions that our creative minds conceive are tempered by our societal upbringing. The weighing of the trade-offs we make is similarly biased. Hence the optimum solution for one person may not be the optimum solution for another.

Consider the development of the heart pacemaker. This electronic device is implanted in the body to regulate the heartbeat, should it become irregular. Before engineers could proceed very

Table 1.1 Comparison between different fields of thought

Engineering	Science	Mathematics	Social sciences and humanities
Study of human-made world	Study of natural world	Study of mathematical constructs	Study of human mind and perception
Engineering design	Scientific inquiry	Mathematical analysis	Rhetoric and criticism
Iterative design process, optimum solution	Hypothesis testing and evaluation	Theorems, proofs, rational constructs	Eclectic methods, comparative values
Biotechnology (heart pacemaker)	Biology (physiology)	Sinusoids (modeling heart)	Ethics (who receives assistance)

far with the design, they or someone on their design team needed to understand human physiology. What materials could be implanted into the body? What was a normal heartbeat? The heartbeat pattern is important, and mathematical modeling is a technique engineers use to assist them in this determination. For instance, a heartbeat, shown in Figure 1.1, may be modeled as the sum of sinusoidal waves added through superposition.

Technology begets change, often change for the better, as in the case of the pacemaker. But what if the technology is expensive? How will it be determined who should receive the aid? In the United States, two-thirds of the patients receiving kidney dialysis are over 65 years old, while in the Great Britain, no one over the age of 55 receives kidney dialysis treatment as part of the national medical program. Technology can also create vexing problems for which there are no ready solutions. Engineers' voices are needed if wise solutions, balancing different views, are to be found. Thus far, engineers have not been well heard in the halls of Congress. The 103d Congress had one senator (1/100) with an engineering or science background and only eight representatives (8/435).

The concept of best or optimum is one familiar to us all. For instance, when you purchase goods or services, an implicit balancing is going on in your mind. Consider that you want to buy the best potato chip. What are the attributes of potato chips that you think are important? Your list will probably include crispiness, taste, saltiness, greasiness, cost, thickness, size, aroma, packaging, and dippability. Somehow you balance all these factors and reach a conclusion as to which chip is the best for you, recognizing that your brother, sister, parents, and friends may reach different conclusions based on their preferences. An analogous weighing of factors, making trade-offs, is used in the design process to create the optimum solution.

Modern engineering is a fairly new discipline, extending back only as far as the mid-1800s. Of course, companies built machines, important ones, before this time, but the understanding of the laws of science and tools of mathematics was very rudimentary and sometimes totally in error (e.g., the caloric theory of heat). The engineering that you will be experiencing is a blend of mathematical analysis, scientific inquiry, and creative design, built on its centuries-old craft origins. The laws of mechanics were well understood by the mid-1800s, as were the fundamentals of electrical engineering. By the end the 1800s most of the natural laws, such as Maxwell's equations describing electromagnetic waves, were discovered and formulated. Universities began incorporating science and mathematics in engineering studies by the late 1800s, a system that has been refined to the present.

An interesting, albeit troubling, aspect of most history and social studies courses in grades K through 16 is the lack of content on technology and its impact on society. People often admire scientific and technological advances and become quite enamored with

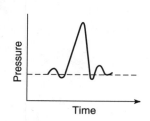

Figure 1.1
Heartbeat.

the gains (e.g., the camcorder, videocassette recorder, personal computer), yet remain ignorant of the processes and knowledge required to effect such gains. When C. P. Snow discussed the separation of the world of science from that of humanities, he discovered that the literary group (those whose schooling and experience fall under the social sciences and humanities heading of Table 1.1) were in many ways much more ignorant of science and technology than scientists and technologists were of literature. You can verify this by reflecting on your own educational background—what portion of your schooling looked at technology, its creation, and its use? This was not always the case; the papers of Benjamin Franklin and Thomas Jefferson reflect an interest in and knowledge of science and technology and its necessity for developing a stable nation. It is unfortunate that the emphasis was lost. Study of the scientific and technological enterprise should not remain the domain of scientists and engineers (if it is addressed at all), but should be included in everyone's education.

A History of Engineering Innovation

We live in a modern, technologically sophisticated society because of myriad scientific and technological advances that have occurred over thousands of years. Figure 1.2 illustrates a time line for some technological advances that have brought us to where we are today. It is difficult to imagine life without many of them. We will examine some of them to gain insight into the leap of imagination and creativity that engineers have exhibited throughout the course of human history. Notice the tremendous advance in the 19th and 20th centuries as engineering evolved into its modern discipline, fusing craft and creativity with scientific understanding and mathematical analysis.

As we all know, people need food, shelter, and clothing to survive. The first inventions, as people advanced from being hunter-gatherers to living in agriculturally based societies, were ones that facilitated this. The yoke and scratch plow enabled animals to be used in the tilling of land, expanding the acreage on which a person could grow crops and thus increasing the number of people that could be fed. The yoke for oxen, created over 5000 years ago, is still used today (Fig. 1.3). This is a fine example of bioengineering, where the device fits the animal's physiology, permitting the strength of the animal, acting through its chest to the yoke, to be transmitted via a shaft to the plow.

With the increase in societal complexity as civilization advanced, the need for standardized weights and measures became important, a concept that we accept automatically today but one which was a major philosophical advance for that era. How does one establish an equivalency for an amount of cotton versus grain versus metal? We know there is no conservation law for volume, but there is a conservation law for mass. Balance-beam scales

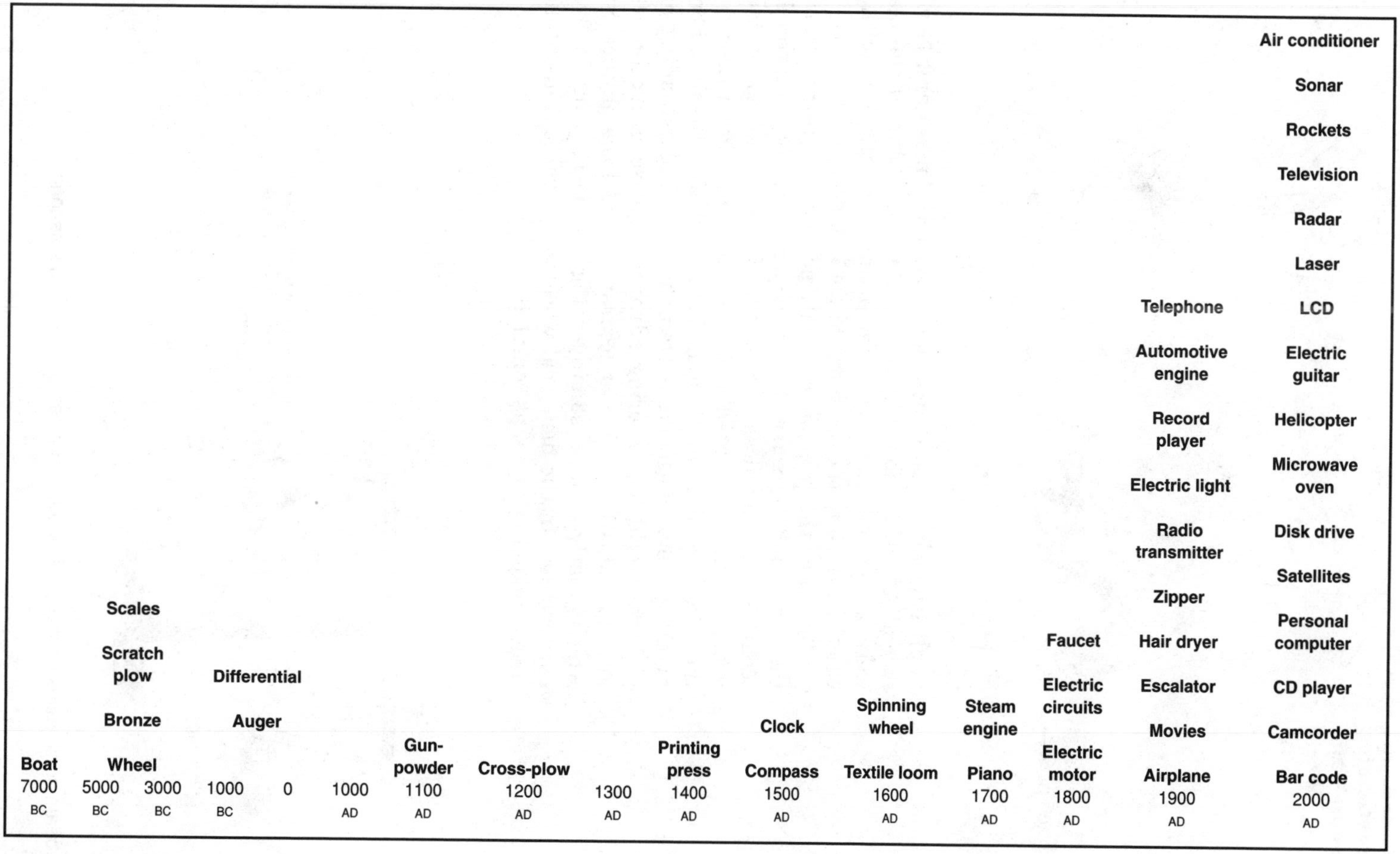

Figure 1.2
Time line of technological advances.

Date	Technological advances
7000 BC	Boat
5000 BC	Wheel, Scratch plow
3000 BC	Bronze, Scales
1000 BC	Auger, Differential
0	
1000 AD	
1100 AD	Gunpowder
1200 AD	Cross-plow
1300 AD	
1400 AD	Printing press
1500 AD	Compass, Clock
1600 AD	Textile loom, Spinning wheel
1700 AD	Piano, Steam engine
1800 AD	Electric motor, Electric circuits, Faucet
1900 AD	Airplane, Movies, Escalator, Hair dryer, Zipper, Radio transmitter, Electric light, Record player, Automotive engine, Telephone
2000 AD	Bar code, Camcorder, CD player, Personal computer, Satellites, Disk drive, Microwave oven, Helicopter, Electric guitar, LCD, Laser, Radar, Television, Rockets, Sonar, Air conditioner

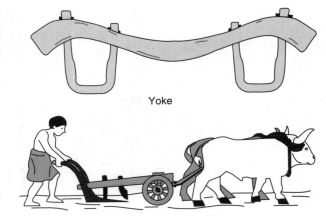

Yoke

Figure 1.3
A yoke and scratch plow—
circa 3000 BC.

implicitly rely on two laws: conservation of mass and Newton's second law of motion. Figure 1.4 illustrates this. If the masses were not equal, then the forces acting upon them could not be equal either, and the scale would not balance.

Note that the previously mentioned laws were not known at the time and the scientific principles were not discovered, yet technological advances were made. It is true for many technological advances created over the course of human history that the science explaining them is discovered after the fact. Perhaps in today's era, with our extremely deep understanding of the physical world, engineering advances and scientific knowledge are coupled. We are able to accurately model the physical world, which greatly assists us in inventing new designs and innovating earlier ones. In the biological world, our level of understanding, hence modeling, is not as complete.

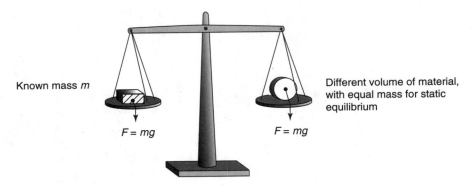

Known mass m

Different volume of material, with equal mass for static equilibrium

$F = mg$ $F = mg$

Figure 1.4
The balance-beam scale relies on the conservation of mass and Newton's second law of motion.

Until 3000 to 4000 BC, humans relied on wood, bone, or rock to shape their environment. Then rudimentary metallurgy developed with pit-fired copper. Copper ore was placed in the bottom of a pit and covered by charcoal. The high temperature of the fire allowed the reaction of the oxide in the copper oxide ore with charcoal, leaving pure metal and a slag at the bottom of the pit. The slag was chipped away, and the copper was remelted and then formed into various shapes or poured into molds.

There is an interesting theory that pottery led humans to discover copper and later, bronze. Clay pots were placed in pits and covered with charcoal so the fire would harden the clay. The glaze used to decorate the pots in all likelihood contained copper or tin oxides. These same oxides (ores) were later used to make copper and then the first alloy, bronze, which is a combination of copper and tin.

In today's world, sometimes referred to as the information age, news of an invention can be disseminated quite quickly and accurately. Of course, this was not the case thousands, even hundreds, of years ago. Inventions could be created and then disappear from the human knowledge base. The Romans discovered cement and used it in the construction of their roads. With the collapse of the Roman Empire, this knowledge was lost for about 1500 years. Inventions are not the domain of the Western world, but exist

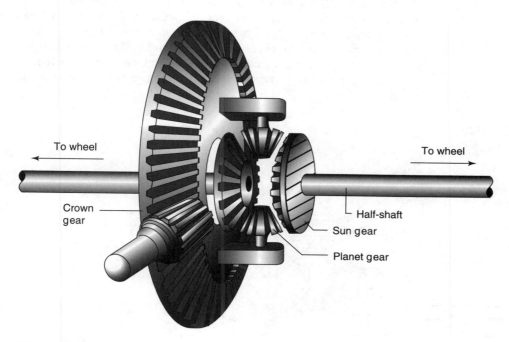

To wheel

To wheel

Crown gear

Half-shaft

Sun gear

Planet gear

Figure 1.5
A differential gear allows wheels to rotate at different speeds, such as when going around a corner.

wherever people developed societies. For instance, the first differential gears (see Fig. 1.5) are attributed to an unknown engineer from China. The gears had to be carefully and painstakingly handmade, as there were no machines to perform this task.

Paralleling the creation of the scratch plow in Egyptian and Mesopotamian times (circa 3000 BC), the cross-plow or moldboard plow (Fig. 1.6) aided in the agricultural development of Europe 4000 years later. The European soil was heavier and had thick grasses that had to be cut through, a task the scratch plow could not perform. The cross-plow, originally pulled by teams of oxen, solved this problem by cutting through the soil and turning it over. With the expansion of agricultural efficiency, the population of Europe increased dramatically, setting the stage in several hundred years for the industrial age.

Many devices were created, as the industrial age evolved, as the basis for societal organization, with three providing the greatest impact—the printing press, the clock, and the steam engine. The printing press (Fig. 1.7) allowed the dissemination of information without personal contact. Prior to its creation, scribes wrote manuscripts by hand, a time-consuming, expensive, and error-prone process. Other information was delivered orally by messengers, which was also an error-filled process. Paper manufacture had made its way via China and the Middle East to Europe by the 1300s, providing an inexpensive medium, compared to parchment, for printing. A type of printing press which has been in use for hundreds of years employed carved blocks of wood that were placed in the press and paper; and it stamped a design on paper, parchment, or cloth. Various factors and knowledge systems coalesced to provide the timber for the creative spark to ignite.

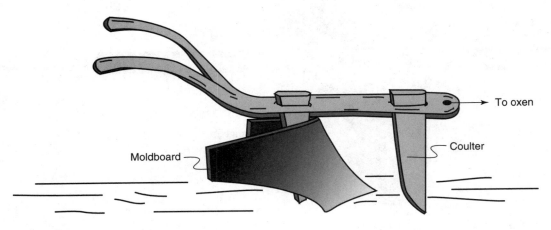

Figure 1.6
Cross-plow or moldboard plow.

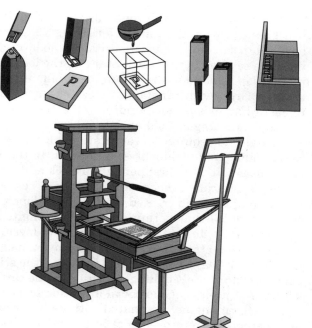

Figure 1.7
The Gutenberg printing press.

In about 1450 Johannes Gutenberg, with a knowledge base in metal casting, actually devised the uniform, movable metal type necessary to create the printing press. The letters had to be of uniform size so the spacing of words would also be consistent. They had to be reusable and had to be firmly held in place while the printing was going on. Gutenberg used a flexible iron band to hold the type in place. The paper was laid on top of the frame holding the inked type and was pressed against the type with a screw jack. The paper adsorbed the inked impression, and a page was printed. Gutenberg had knowledge of metal processing; without it he could not have created the movable metal type. Your engineering education provides you with similar knowledge bases in a variety of technical areas. Your use of this knowledge and your natural creativity will allow new and meaningful technological designs to evolve.

At the dawn of the industrial era, travel by land was very time-consuming, as the roads were made of dirt and easily became rutted. Water travel provided an attractive alternative, and ships were widely used whenever possible. The difficulty for seafarers lay in knowing where they were, if land was not in sight. The development of the clock, initially motivated by astronomical reasons, helped expand this industry and, with it, travel around the globe, dramatically opening trade with the Far East and Europe. Caravans had crossed from the Middle East to China and back for hundreds, if not thousands, of years, but the amount of commerce that reached Europe was scant. The demand for goods was great, providing incentive for change. Figure 1.8 illustrates how a ship can determine its position—the latitude—by knowing its angle relative to the North Star and the longitude by knowing its location relative to the zeroth meridian, or that which passes through Greenwich, England. Tables of stars' positions relative to earth had existed for hundreds of years, and with the aid of a sextant, the angle could be determined and hence the latitude. The longitude required an accurate clock which would work at sea, a rolling and turbulent environment. In the 1650s Christian Huygens created the first pendulum clock and a mechanism that provided accurate timekeeping. It took a century before John Harrison of England created the first chronometer (an extremely accurate clock) that permitted accurate longitude location. For an interesting story of the chronometer's development, and the obstacles this creative genius had to overcome, read the book *Longitude* by Dava Sobel. In this instance, the science and mathematics of accurately finding a position on the earth's surface were known, but it took technological invention to solve the problem.

The steam engine, associated with James Watt, who actually held patents on improvements in steam engines, was originally designed by Thomas Savery and Thomas Newcomen. Watt created two significant improvements: (1) an external condenser for the steam, so the steam cylinder would not be cooled down, and (2) a slide valve mechanism that allowed steam to automatically enter the cylinder as the piston was pushed by the steam in either direction. These fundamental inventions were not enough to garner success in the marketplace. In this era, the mid-1700s in England, the designer of the engine would have to manufacture and install it at a factory and then be compensated for the effort. This took more capital than Watt possessed, and he sold one-half of his patent to a partner, John Roebuck. Together they were able

Figure 1.8
The ship at *B* can determine its latitude based on its incident angle to the North Star. The earth rotates through 15° of longitude each hour (360°/24 h = 15°/h). If the local time is known, say, 12 noon, and the time in Greenwich (0° longitude) is known, say 3 PM, then the local longitude is 45° east of Greenwich. Noontime is often used as that time is when the sun is overhead. An accurate clock, a chronometer, is necessary as even a few seconds' error results in many miles in error.

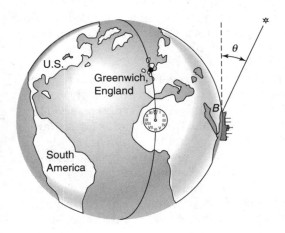

to build several engines, but the financial return and Roebuck's capital were limited, so the business could not expand. Time was also running out on Watt's patent. Fortunately Matthew Boulton, with sound financial resources, took an interest in the engine and purchased Roebuck's share. Boulton also had the connections to get the patent extended for another 17 years, securing his investment. Because of Boulton's "deep pockets," Boulton-Watt Company created an interesting and profitable marketing tool: It would provide an engine free to the purchaser in exchange for only 10 percent of the savings that accrued for its use per year for as long as the engine was operating. Engines from other manufacturers, often exhausting to the atmosphere, were comparatively very inefficient, so the manifold efficiency improvement of the Watt engine provided the purchaser with significant savings. Ten percent of these funds, also significant, flowed to Boulton and Watt. This story illustrates an important lesson: Inventiveness is important, but seldom does one person have all the technical, marketing, and financial know-how required to make a financially successful product. A team effort is required.

Steam was not the only form of energy of interest at the time. The study of electricity was in its infancy, and great strides toward understanding it were being made, though not always by design. In the late 1700s, Luigi Galvani, a professor of physiology, found that touching a frog's leg with metal would cause the muscles to twitch. This led to two conclusions: Animals, hence people, generated electricity, and muscles were activated by these generated electric signals. About 10 years later, Alessandro Volta, a professor of physics at the Royal School in Como, Italy, showed that the electric current causing the muscles to twitch in Galvani's experiment was the result of its being generated by dissimilar metals (galvanic action, as it is currently known). He went on to construct the first battery, known as the voltaic pile, by alternating strips of copper and zinc in slightly acidic water.

Advances in one area, the voltaic pile, allowed other scientists to pursue related interests, namely, the connection between magnetism and electricity. In 1820 a Danish physicist and chemist, H. C. Oersted, set out to prove that there was no connection between the two phenomena but discovered, to the contrary, that they are intrinsically connected. He showed that electric current flowing in a wire sets up a magnetic field around it (Fig. 1.9) and he demonstrated this by a compass needle moving and aligning itself with the poles of the generated field.

Maintaining an open mind is crucial to the creative and investigative process. Looking for only one type of solution can block you from seeing others that are viable. There are numerous anecdotal stories, such as Oersted's, illustrating this.

Another pattern emerging from studying the advances in science and technology is one of reliance on others. In other words,

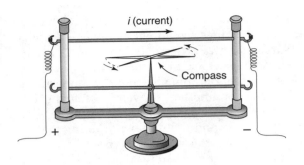

Figure 1.9
Electric current flowing
through a wire creates a
magnetic field around it.

the use of the knowledge gained and disseminated facilitates further advances. One does not have to start at the beginning, but one should creatively use prior gains. The development of the telephone illustrates this, as Alexander Graham Bell used the works of predecessors together with his own innate creativity to devise the first telephone. Bell was interested in communication, particularly in helping deaf people communicate. While attending the Massachusetts Institute of Technology, he heard a lecturer who showed that differing sound waves produced unique visual patterns. He demonstrated this by using a membrane with an attached bristle brush to scribe. Bell was initially interested in the visual portrayal of sound. This did not prove fruitful, but he later connected the principle of a vibrating membrane to the way electric energy is created. The device he made (Fig. 1.10) shows that when the pressure waves generated by the voice activate a membrane connected to an electromagnet which vibrates, electricity is sent through a wire to a second electromagnet that in turn vibrates, causing its connected membrane to vibrate and generate pressure waves and sounds that are heard by the listener.

Thomas A. Edison is the person whom almost everyone has heard of in relation to creativity and the design of devices. He invented the phonograph (1878) and the electric lightbulb (1879) and has over a thousand other patents applied for in his name. Prior to his invention of the lightbulb, scientists knew that carbon would glow when heated to a high temperature by the passage of electric current through it. The problem for lightbulbs was that the carbon element quickly oxidized. Edison and his staff

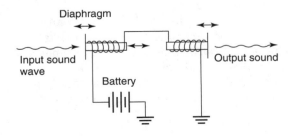

Figure 1.10
Schematic diagram of
Alexander Graham Bell's
telephone.

investigated the properties of over 3000 different materials before deciding that carbon was the optimum material. They were able to create a very high vacuum in the glass bulb so the oxidation process would be slow, devising a bulb in 1879 that lasted for 170 hours (the lightbulbs of today use tungsten filaments and are filled with an inert gas, argon, to prevent oxidation).

The phonograph and electric lightbulb provided a tremendous amount of income to Edison, funding additional investigations. He is credited with conceiving of the first "invention factory," a research and development facility that turned advances in scientific research into marketable products. One of Edison's maxims was that any idea he developed should have the potential to be profitable. He used market analysis as a check before deciding whether to pursue an invention; but once he did decide, he was tenacious. The long hours he and his staff worked are legendary—90 to 100 hours per week while in the pursuit of a solution. His work demonstrates the need for dedication, the requirement of teamwork with differing areas of expertise, and the importance of marketing and sales if inventions are to be profitable.

There are many other interesting stories that illuminate the path of industrial development in the 19th and 20th centuries. As one last and more current example, look at the bar code. Bar codes provide a numerical way to define a product or location (e.g., postal zip codes and food product codes). The food product is identified by a number which is then converted to binary form and read by an optical scanner. The postal bar code provides information about the street address that can be accurately and automatically read by a scanner. Because of scanners, the U.S. Postal Service is able to improve productivity and reduce costs. Based on 1990 costs, the productivity of the U.S. postal worker, measured in pieces of mail handled per employee, was more than 3 times that of the Federal Republic of Germany; and the cost of a first-class letter was about $2\frac{1}{2}$ times less than that in Germany.

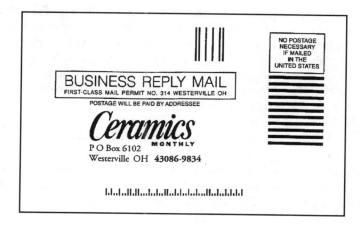

Figure 1.11
Postcard with a bar code.
(*Used with permission of* Ceramics Monthly.)

The post office district is defined by a five-digit number and the street address by a four-digit number, which are again converted to binary form. These are the lines that you find on the bottom of envelopes or postcards, such as in Figure 1.11.

Table 1.2 POSTNET binary code system

Character value	Code
0	11000
1	00011
2	00101
3	00110
4	01001
5	01010
6	01100
7	10001
8	10010
9	10100

$$4 + 3 + 0 + 8 + 6 + 9 + 8 + 3 + 4 = 45$$
$$10 - 5 = 5 \longleftarrow \text{check}$$

Figure 1.12
Converting a POSTNET
bar code to its numeric
value.

The POSTNET binary system is shown in Table 1.2. Using this, we can convert the symbol in Figure 1.11 to its numerical value. The long bars are 1s, the short bars are 0s. The POSTNET code always begins with a long bar, telling the system to begin, and ends with a long bar, telling the system to stop. In addition, there is a check built into the code. You will notice in Figure 1.12 there is an extra number. This serves as a check. Add all the numbers in the zip code, and you get 45. If you consider only the digit in the 1s place, and subtract it from 10, you get 5.

The Nature of Science, Mathematics, and Engineering

Engineering practice requires the use of science, mathematics, and social sciences and humanities for its successful execution. There are values associated with each of these areas, values that are implicit in the maxims of each. For instance, science has an underlying belief that the world is understandable. This extends beyond our immediate world to the universe; and the laws (e.g., gravitational attraction) that determine behavior on earth are also valid for the moon, Mars, and faraway galaxies. While we cannot ever hope to achieve perfection in complete understanding and hence resolution of all questions, the quest for understanding is a worthy endeavor. We know that the ideas of science may change with time. For instance, the Copernican view that the sun rotated around the earth changed, and ideas concerning how heat was transmitted changed as new evidence was discovered.

Science relies on evidence, hypothesis, and logic. However, the practice of science may differ in what is viewed as important to investigate, the qualitative and quantitative methods used, and the conclusions drawn. Evidence is paramount to the scientific process; hypotheses without evidence are merely assertions. The observation of phenomena validates a theory, an explanation, or it will not. Often great skill is needed in making the observation so as not to have it disturb the phenomena, an extremely difficult problem when one is observing living organisms. Other difficulties can arise when the phenomenon, perhaps the growth of trees in forests, has many factors affecting it, and distinguishing between multiple effects is challenging. Logical reasoning assists the scientist in validating hypotheses; the primary cause for scientific arguments concerns the appropriateness of the evidence. Do the global temperature variations indicate global warming, or are they part of the cyclic nature of temperature change that has

Scientists checking the Relativistic Heavy Ion Collider (RHIC) at Brookhaven National Laboratory. The RHIC attempts to recreate the conditions at the beginning of the universe. It accelerates heavy ions, such as gold, and then has them collide with each other, reaching an energy density close to that of the Big Bang, a theory of how the universe began. (*Courtesy of Brookhaven National Laboratory.*)

occurred on earth? How does one separate the effects? There will be certain assumptions, and these assumptions are debatable; but based on the assumptions, a logical train of thought will lead to a conclusion. One of the ways that theories are validated is through their use in explaining and predicting events.

The scientific method tries to be bias-free, recognizing that we may bring certain biases to the observation of data that we are not aware of. For instance, in studying the behavior of primates by male scientists, the major focus was on the competitive social behavior of male primates. When female scientists entered the field, the importance of female primates in community building came to light. A recognition that bias based on one's background may cause one to look for certain types of evidence and not others, or interpret evidence from that perspective, is something to be aware of. Logical persuasion assists tremendously in minimizing these risks.

Mathematics is used as a tool by scientists and engineers. It shares many patterns with science in that both seek to discover general patterns and relationships. Mathematics diverges from science because it is not bound by nature, but may explore abstractions having no real-world connections. Of course, there are many applications for mathematics, from music to engineering to business. As indicated earlier in the chapter, mathematics allows us to express ideas (Newton's second law) in symbolic form and then manipulate this form. A challenge for the engineer and scientist is to make certain that the model they are using is consistent with mathematical assumptions. For instance, the properties of wood vary depending on the grain direction. Assuming that the properties are uniform in all directions may not yield accurate results.

Assigning physical values to symbols is important and does influence the mathematics. For instance, $3 + 2 = 5$, but 3 apples plus 2 oranges does not equal 5 apples. When 2 cups of coffee is added to 3 cups of coffee, the pot will contain 5 cups of coffee. When 2 cups of sugar is added to 3 cups of coffee, the pot does not contain 5 cups of coffee. In the latter case, there is no conservation of volume law, while the law of conservation of mass underlies the addition of differing amounts of coffee.

Technology, created through engineering design, enables us to change the world to better meet our needs. The results can include unexpected benefits as well as unforeseen costs and risks. Engineering blends the practical knowledge of fabricating devices with the scientific knowledge of why they perform, to create new technologies. Reflect on engineers such as Alexander Bell or Thomas Edison, and note the integration of theoretical and practical knowledge in creating new technologies. Technology and science are inexorably linked; one assists the other and vice versa; the development of the computer has expanded the study of weather systems. Concepts such as the first and second laws of thermodynamics were linked to understanding and improving the steam engine. As technologies become more sophisticated, the division between engineering and science becomes blurred, difficult to observe. In the study and design of solid-state devices, physics and engineering are intertwined. Technological advances require new understandings, and scientific investigations often require new technologies.

Constraints

The engineered world interacts with social and cultural values more directly and immediately than scientific or mathematical worlds do. Technologies can create benefit and risk, so trade-offs or constraints are part of the process. There are certain constraints, such as the conservation of energy, even the conversion of energy, that are dictated by natural laws. We cannot create an engine with 100 percent efficiency; it is theoretically impossible. These constraints are understood because of engineers' background in science; but the constraints in which a balance must be sought between economics (limited time and material), politics (regulations), society (public opposition), and ethics (disadvantaging some people) are not analytic.

All these factors come into play when a shopping mall is to be constructed. The developer wants to spend only a certain amount of money building a mall so that it will be profitable (the developer defines appropriate profit, generally); there are regulations regarding building size, number of parking lot spaces, and traffic

Engineers checking drawings of an off-shore oil rig. (*Courtesy of Texaco, Inc.*)

patterns that need to be satisfied; abutting property owners may protest; people may be concerned about the environmental impacts and the aesthetic impact on the community; many local merchants may face possible business loss, while other people may find employment opportunities. Thus, a final design, be it a construction project or an aircraft design, involves many trade-offs, yielding an optimum solution. The optimum solution may change if new technologies appear or economic constraints change. We will also discuss the creative part of the design process, the portion that is not logical, but intuitive and imaginative, an important part of engineering.

Impact of Technology

We have seen that new technologies are created to solve a human problem or address a human need, but frequently the technology brings unanticipated consequences as well. At times this occurs because our vision in applying the technology is limited. The application of pesticides to lawns and gardens, including vast acreages of crops, increases the yield of fruits and vegetables, a desirable feature. The pesticides leach into the soil, killing some good organisms, and then percolate farther downward into the water supply, causing eventual contamination, an unforeseen, though predictable, negative feature. In the 1920s there was little refrigeration, food spoiled quickly and people were made critically ill because of this. The refrigerator, using dichlorodifluoro methane, was invented and widely adopted, improving the quality of food that people could purchase and later safely keep in their homes. Fifty years later, scientists discovered that this hydrocarbon is greatly contributing to the depletion of the ozone layer in the earth's upper atmosphere. Its use has been banned in refrigerating and air conditioning equipment in most of the industrialized world. A complicating effect is that currently the smuggling of this refrigerant into the United States is highly lucrative, and a black market flourishes in some areas. Equipment that used the original refrigerant cannot use the replacement refrigerant, so there is a great economic incentive to continue using it until the device cannot operate anymore.

The computer has revolutionized many aspects of life in the industrialized world; ever more frequently people operate computers that control machines that make a product. In the pre-industrial world, craftspeople made the product by hand; as factories evolved with industrialized society, people operated machines; and now people operate computers. Reporters use computers to write their stories, often sending the stories directly to the presses; however, reporters also indicate eyestrain and repetitive stress syndrome in their hands and arms from keyboarding in one position for long periods, resulting in loss of work and sometimes permanent disability. The impacts have moved from catastrophic, such as losing a limb in machinery, to chronic, such as back pain or carpal tunnel syndrome. Employment has improved for people making computers; there is an increased demand, and they need to be repaired. But the printer who set the type from the copy provided by the reporter no longer has a job, another unintended consequence.

Coupled with the realization that there will be unintended consequences in the human-made world, there is the realization that the device or system will fail at some point. Failure is natural; it occurs when living organisms die, when physical devices cease to function. The more complex the system, the more ways there are for it to fail and the greater the challenge to prevent unanticipated failure. There are various strategies for preventing failure, including overdesign—making the device bigger or stronger—and backup systems that will operate when the primary system fails. No device is fail-safe, almost fail-safe, yes. Methods of risk analysis, including the probability of undesirable occurrences, can be used in the design process of complex systems, or systems that may have dramatic negative consequences. Such systems are often associated with nuclear power plants and hazardous material manufacturing or processing facilities, and may be included as part of the design process when possible danger to the public or workers exists.

Fields of Engineering

There are many fields of engineering and specialties within fields and across fields. For instance, a person may be an electrical engineer specializing in engine control systems or a mechanical engineer specializing in engine control systems. Frequently engineers with a primary focus in one discipline require knowledge from other disciplines to complete a task. In large companies a team of engineers with different backgrounds will work together on a project; in smaller firms, an engineer must develop sufficient expertise to solve the interdisciplinary problem. If the problem is too complex, consultants may be hired. One reason for studying certain common engineering courses, say, materials science, electric circuits, thermodynamics, and mechanics, is to develop a broad basic knowledge of engineering necessary to solve the complex and interdisciplinary problems typically confronting engineers in practice.

The following engineering fields are by no means comprehensive, but represent the majors most frequently found at universities. The descriptions are necessarily brief sketches with some key distinguishing attributes. For expanded discussion, please refer to the Web sites of the professional societies. You will glean a great deal of information in doing so. Check out several sites that may be of interest; notice the breadth of activities and areas of interest in each of the disciplines.

Aeronautical and Aerospace Engineering

Aerospace engineering comes to mind when we think of aircraft, rockets, and space travel. Actually, these are incredibly complex interdisciplinary systems requiring the talents of engineers from many fields of study. However, aeronautical engineering concerns itself with flight and the movement of fluids in the earth's atmosphere. Aircraft in this atmosphere can use oxygen for combustion and other processes; not so for aerospace devices traveling beyond the earth's atmosphere. In the design of aircraft, several factors need simultaneous consideration—aerodynamics, propulsion, controls, and structures. You may have noticed the flexing of an airplane's wings as it flies through the air, a natural occurrence resulting from the interaction between the air and the structure. It is an interaction that the design accommodates. Aerospace engineers launch people and equipment into space with rockets. They must consider the aspects of aircraft design, but the applications and constraints are different. www.aiaa.org

Bioengineering

Bioengineering is a comparatively a new field of engineering that began in the 1940s but has evolved for thousands of years, since the first artificial limb was created. This is an interdisciplinary

Students checking the pressure drop around an airfoil in a subsonic wind tunnel. (*Courtesy of Hofstra University.*)

field, as bioengineers must be able to work and communicate with physicians and biologists in developing new equipment and materials. When discussing the development of the pacemaker earlier in this chapter, we noted that engineers needed to understand the immune system (so materials would not be rejected by the body), how the heart worked and the level of electrical stimulus needed, the effect of the power source on the body, and how the physician could perform the operation. In gathering these data, bioengineers face the challenge of determining information from living organisms and developing noninvasive procedures to prevent measuring the reaction to the measuring process. Very often this requires a graduate degree because of the diverse nature of the material that must be comprehended. www.mecca.org/bme/bmes/bmeshome.html

Chemical Engineering

The scientific breakthroughs in a chemistry laboratory must be translated to commercial realities by chemical engineers. These engineers work in the pharmaceutical, chemical, nuclear, and electronics industries. They design the processes, specifying the equipment that will solve particular production problems. Fundamental to designing equipment for these processes is a knowledge of the conservation laws of mass and energy, chemical equilibrium, and reactivity. Most processes are continuous, so a thorough understanding of automatic control systems is important. A new pharmaceutical, Xmycin, has been developed by biochemists; chemical engineers must translate the laboratory processes to a pilot plant. Problems of scaling up the processes must be analyzed and overcome, and the equipment designed and controlled. If the pilot plant operation is successful, then the process repeats itself for full-scale production. The scaling of processes is nonlinear, much as a food recipe for 50 people cannot be reduced to one for 2 people in proportional fashion. www.aiche.org

Civil Engineering

Civil engineering is perhaps the most apparent field of engineering to us. The highways and bridges we drive on, the tunnels we drive through, the buildings we live in, and the water and sewage treatment systems we depend on are the handiwork of civil engineers. To be sure, as civil engineers you will need a sound understanding of mechanics, materials, and structures as well as soil properties. If you are required to build a structure on damp soil, how can the load-bearing properties be enhanced? Many structures at LaGuardia Airport in New York City are built on wet, sandy soils. Pilings are driven into the soil to provide stability to the structures built upon them.

Highway construction, indeed the construction of all large systems, requires accurate surveying of the terrain, defining aspects of the terrain that require consideration in the construction process. Because civil engineers design and construct systems (transportation, water) and products (dams, buildings) that are so vital to society, most civil engineers are licensed professional engineers, more so than engineers in any other field. Very often civil engineers work for local and state governments and frequently also act as city planners. www.asce.org

Computer Engineering and Computer Science

Computer engineering is closely connected to computer science, and both are involved with the design and organization of computers, their software, and their hardware. Computer science approaches the design and organization from the software view with less emphasis on hardware, and computer engineering from the hardware view with less emphasis on software. The computer engineer is concerned with how to take the software commands and translate them to electronic signals within the computer and within computer systems. Simply because a software command says to do something, how does it physically happen? You push the letter a on the keyboard, but how does it appear on the screen? Following this relatively direct process highlights the difference between computer scientists and computer engineers. www.acm.org

Project engineer overseeing the proper installation of an underground conduit. (*Courtesy of Sidney Bowne and Son.*)

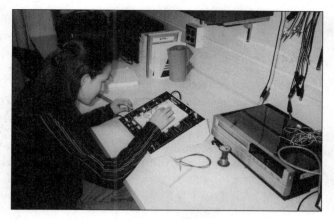

Student designing bioinstrumentation amplifiers to improve the signal-to-noise ratio in an electrocardiogram recording. (*Courtesy of Hofstra University.*)

Electrical and Electronics Engineering

Electrical engineering deals with the motion of electrons in metals, such as through wire and filaments, while electronics deals with the transmission of charged particles in a gas, vacuum, or semiconductor. Just imagine a world without electricity. No lights, elevators, refrigerators, motors, dishwashers, subways, or televisions. Our life is integrally linked to electrical systems and devices, all designed and analyzed by electrical engineers. The growth of electrical engineering has been exponential as microprocessor-based and computer-based systems have become common to our everyday life—microwave ovens, computers, cellular phones. Some systems are so complex that humans cannot control them manually; for example, the forward-swept-wing fighter planes are unstable unless they are computer-controlled. All electronic or electrical equipment requires the knowledge that an electrical engineer possesses to be designed, developed, and manufactured. www.ieee.org

Environmental Engineering

Environmental engineering traditionally developed as a specialization within civil engineering and is now assuming greater independence as a field of its own. It is a field that crosses the boundaries of several other disciplines. All engineers are concerned with the environment and create processes and products that minimally disrupt the natural environment. Environmental engineers may be chemical engineers focusing on the containment of environmentally hazardous materials, or mechanical engineers concerned about pollution from combustion processes, or civil engineers looking at waste disposal problems. Not only must the environment be assessed for levels of pollution and contamination, but also engineers must create systems to remediate the contamination and restore the natural environment. If soil contamination is found at an old manufacturing facility, the soil must be restored to environmentally safe conditions before the facility may be used for new functions. Environmental engineers will design processes to do this. www.wef.org

Industrial Engineering

Industrial engineers are concerned with the design, improvement, and installation of integrated systems of people, materials, and energy. Consider the manufacture of tires: A machine is operated by a person and requires certain raw materials to produce the tire. The problem for the industrial engineer is to bring the material to the machine as it is needed, remove the finished product, and transport it to the next workstation. When establishing a manufacturing procedure, the industrial engineer considers the machine's capacity, when routine maintenance is needed, and what the operator should do when the machine is not functioning. The industrial engineer is always focused on improving quality and reducing costs; hence a curriculum includes courses in economics and finance, industrial psychology, as well as engineering fundamentals. www.iienet.org

Mechanical Engineering

Mechanical engineers apply the principles of mechanics and energy to the design of machines and devices. Applied mechanics is the study of motion and the effect of external forces on this motion, so the mechanical engineer is involved with crankshaft design and turbine rotor design. Engineers in this area must consider the vibration effects of the device on the system and counter the situation with vibrations imposed on the device. In designing the nozzle of a rocket, the engineer looks at the design from two viewpoints: the fluid's effect on the nozzle and the nozzle's effect on the fluid. There is a keen interest in the interactions between the fluid and the solid interface, such as in the imparting of fluid energy to turbine blades to produce power. Of course, material selection is fundamental to a good design, and engineers must be aware of how strong the material is, how it can be lubricated if it moves, and how it will wear. www.asme.org

References

1. Burke, James. *Connections.* Boston: Little, Brown, 1978.
2. Florman, Samuel. *The Existential Pleasures of Engineering,* vol. 1, 2nd ed. New York: St. Martin's Press, 1994.
3. Landis, Raymond C. *Studying Engineering: A Road Map to a Rewarding Career.* Burbank, CA: Discovery Press, 1996.
4. Marchaj, C. A. *Sailing Theory and Practice,* New York: Dodd, Mead, 1964.
5. Snow, C. P. *The Two Cultures.* New York: Cambridge University Press, 1993.

6. Sobel, Dava. *Longitude: The True Story of a Lone Genius Who Solved the Greatest Scientific Problem of His Time.* New York: Walker and Co., 1995.

7. Tenner, Edward. *Why Things Bite Back—Technology and the Revenge of Unintended Consequences.* New York: Knopf, 1997.

8. Volti, R. *Society and Technological Change,* 3rd ed. New York: St. Martin's Press, 1995.

9. *Science for All Americans.* Project 2061, American Association for the Advancement of Science. New York: Oxford University Press, 1990.

Problems

1.1. Determine the temperature on your campus. A team of students will be needed to measure the temperature at various locations at the same time. How do you resolve the different readings? How many readings should you take?

1.2. State a hypothesis, such as that more right-handed people have blue eyes than left-handed people do. Set up a procedure to prove or disprove the hypothesis.

1.3. Write an essay (200 to 300 words) on the driving force(s) behind engineering activities in today's world.

1.4. Describe mechanical advantage. How is it manifested in some common household devices?

1.5. Describe how you think the Egyptians constructed the Great Pyramid. The stones on the pyramid weigh about 2½ tons and had to be transported from 500 miles (mi) away.

1.6. Write an essay tracing the development of the steam engine.

1.7. Define *engineering,* and trace the etymology of the word.

1.8. Discuss why you would like to become an engineer. If you are not certain whether you want to practice engineering, discuss what factors will assist you in deciding on an engineering career path.

1.9. Discuss the similarities and differences of civil and mechanical engineering.

1.10. Examine the course requirements for a degree in computer science and for one in electrical engineering. Discuss the similarities and differences.

1.11. Develop a technological time line of your life. Include significant technological events, such as the IBM personal computer introduced in 1983.

1.12. Thomas Edison's lightbulb filament was initially made of extruded carbon, but today the filaments are made from tungsten. Use a reference book to find the electrical resistance of tungsten and of several other metals, such as copper, tin, and iron. Why is tungsten the best choice for a lightbulb?

1.13. Determine the speed of sound. For instance, you could stand a long measured distance, say, 1000 to 2000 feet (ft), from another student. If that student bangs two metal lids together and you measure the time it takes for the sound to reach you, a rough determination may be made.

1.14. Picture phones are telephones that display video pictures of the people talking to each other. Do you think these phones will become commonly used? What effects might the phones have on our lives?

1.15. There has been concern that using cellular phones and living near power lines can both be dangerous because of electromagnetic radiation. Investigate the problem of electromagnetic radiation. Is there reason to be concerned? Are there inherent trade-offs being made in using these technologies that we may not be aware of?

1.16. A process called colorization uses computer technology to turn old black-and-white movies into color ones. Many feel that the movies are enhanced by the process, while others decry this as spoiling the original intent of the filmmaker. What is your point of view on the use of this technology?

1.17. In what ways do you use and/or are you dependent on information that is transmitted by satellite?

1.18. Discuss some of the impacts that the widespread availability of video-cassette recorders has had on your personal lifestyle.

1.19. Communication technology can affect the outcome of military conflict. Investigate the communication tools used in Operation Desert Storm versus the comparative lack of such tools in the Battle of New Orleans during the War of 1812.

1.20. There is increasing concern that the privacy of individuals is being invaded by database companies which sell information to other companies. Is it important for manufacturers to have information about the customers who buy their products? Why or why not?

1.21. Sheet metal is used to make coins through a process of die punching. The metals used for coins are chosen because of their properties such as hardness, ductility, corrosion resistance, and melting point. Using a reference book, find out which of the following metals would make poor choices for coins: copper, magnesium, aluminum, iron, nickel, mercury.

1.22. What types of fuels do you think will replace oil-based fuels like gasoline in the future?

1.23. Should we continue to explore our solar system and the universe? Is the exploration worth the billions of dollars it costs?

1.24. Automobiles give us the freedom to go where we want whenever we wish. Is this freedom worth the increased pollution compared to that generated from public transportation?

1.25. Explain in your own words how an airplane wing allows something that is heavier than air to fly.

1.26. Investigate the development of the first internal combustion engine.

1.27. Describe some design features you would like in your home in 20 years.

1.28. Visit one of the Web sites noted under "Engineering Fields." Mention two aspects of the field that were new to you. What are some of the resources available to members?

2

The Design Process

One of the wonderful aspects of engineering is that it is a creative profession; engineers make new products. Traditionally one associates engineers with being very analytical and good in mathematics and science, and they are; but, equally important, they are creative and must be able to engage this part of their being as well. The requirement for the combination of these two attributes is found in the design process. There are some traits that many creative people share, attributes that can be enhanced if you are aware of their importance.

Traits of Creative People

Engineers creatively solve problems, hence understanding the wellsprings of creativity is valuable to further developing and sustaining it. Creativity springs from the mind, blending our various intelligences together. Design and creativity go hand in hand; enhancing your creativity will improve your designs. The following traits of creative people may appear separable, but they are not.

Taking Risks

Whenever you depart from the norm or the accepted view, you increase your vulnerability. However, increased vulnerability is necessary if you are to be receptive to new ideas. You have no doubt heard the disabling expressions "You shouldn't argue with success" and "If it ain't broke, don't fix it." They imply that if you try to improve something that functions, you will break it. If, however, you are to create something far better than anyone has thought of before, a risk must be taken, a chance to succeed and also a chance to fail.

There are at least two aspects to learning: knowledge gained from successes and knowledge gained from mistakes. Success is very appealing, but there are drawbacks to learning from success, as it tends to focus the direction of thinking toward the future, excluding other views from consideration. Learning from failure, however, does not have an exclusionary view. It is less directed, and alternatives are more readily accepted as possible directions

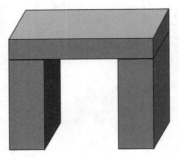

Figure 2.1
A simple arch.

to follow. In learning a new skill, a person must experience periods of awkwardness, a time of inability. Mistakes are made in the process of skill development. Top athletes did not start off at the top; they learned from their mistakes, baskets missed, strikeouts incurred. In creating as well as in learning, this attitude is required.

Vision

What is vision? Perhaps it is seeing connections where none were perceived before, often crossing traditional boundaries between disciplines, looking in a different direction. Changing one's perspective may suddenly allow the perception of interconnections, a bridging of different worlds. One of the ways people achieve this vision is by allowing a certain ambiguity to exist, not seizing a known methodology, precluding other ways of thinking. This manner of thinking links very well to the design process in which several solutions, or alternatives, are sought before they are narrowed down to one.

Our verbal and mathematical intelligences, collectively a symbolic intelligence where we replace meanings with word symbols, cause us to focus on definitions. Thus, when we see the word *elephant,* we do not visualize an elephant, thinking of how it smells, walks, and sounds; rather, we associate a definition with a word, a definition made from other words. This very important type of intelligence can often limit our perception by focusing on the known, the unambiguous. Place three blocks together, as in Figure 2.1. How might the figure be described? Did your listing include a table, a pleasing form, a chair, an inverted *u,* a hole in space, a bridge, the Greek letter *pi*? Visual thinking is an enabling intelligence, empowering you to see in different ways and to make connections between varied perspectives.

Knowledge Base of the Subject

Mastery of a subject or technical ability is a prerequisite for creative thoughts. While it is recognized that acquisition of knowledge brings the accompanying danger of strengthening a fixed view, it also strengthens intuition. The strengthened intuitive knowledge can harmonize the competing tensions within oneself formed by lessons learned from successes and mistakes.

Concentration and Determination

Many people, Thomas Edison included, said that creative leaps are a result of perspiration as much as inspiration, although both are required. The stories of Edison's spending weeks looking through hundreds of references in the effort to turn a conceptual design into a finished design are legendary. The same concentration and drive are necessary in engineering. This does not mean "toughing it out," but rather focusing one's energy through relaxed attention.

The determination to make creative change requires dealing with periods of uncertainty. Creative people have this talent, this ability to cope with their isolation for following a nontraditional path in pursuit of their vision. In the process a resiliency, not a hardening, is created. Resiliency connotes strength and flexibility, whereas hardness connotes strength and rigidity.

Developing Visual Ability

Visual thinking is central to the creative process. This thought process is aided by quick sketching, permitting the ideas to become further refined. We think visually all the time. The English language contains many words linking vision and thought, such as *insight, foresight,* and *hindsight.* Indeed, the etymological root of the word *idea* comes from the Greek word "idein," to see. Thinking is a pervasive, yet elusive, concept. Thinking, as manifested by the brain's electrical activity, occurs even while we sleep. It is a complex activity that involves the entire body—physiologists have shown that muscle tone affects thought; neurologists link the nervous system and psychologists link feelings and emotions to the thought process. The process is visual, verbal, and mathematical and is affected by anything that affects our bodies. Of course, not all thinking is fruitful or purposeful; most of the time it produces little of value, but it is occurring.

Visual thinking pervades all our activities from the profound to the mundane; from Einstein's viewing himself traveling at the speed of light in his search for the theory of relativity, to an athlete's picturing herself winning a medal, to a walk through a crowd of people. How might we interact with our visual thought process? There are three types of imageries that we use: seeing, imagining, and drawing. Figure 2.2 illustrates these as overlapping circles, one reinforcing the other. Drawing does not mean formal drawing, but doodling or sketching. Thus, seeing what is helps us to draw, and drawing what is helps us to see more clearly. Similarly, drawing stimulates imagining, and imagining provides greater desire to draw. Imagining and seeing relate a bit more

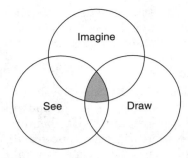

Figure 2.2
A Venn diagram of seeing, imagining, and drawing.

esoterically, as imagining filters what we see and seeing stimulates our imagination. One form of imagery merges continually with the others.

Quick sketches aid the problem formulation and solution by allowing examination from several approaches. Because of the sketching, you do not have to hold different images in your mind, and the sketches stimulate your imagination. This is the flow of imagery from perceptual (seeing) to inner (imagining) to graphical (sketching). Sketching focuses our vague inner perceptions and in the process of recording them, advances them. It also allows for comparisons between different ideas. The word *sketching* is used to denote the concept of graphical ideation rather than the formal process of graphical communication to another person. Our thoughts occur quickly and require an equally rapid graphical format such as freehand sketching.

Right and Left, Left and Right

Folklore tells us that the symbolic left hand represents subjectivity, sensory perception, and subconscious levels of thinking, while the symbolic right hand represents logic, objectivity, skill, and language. In the brain the hemispheric function is reversed. The right hemisphere directs the left side, and the left hemisphere directs the right side of our bodies. Scientists have confirmed the functional aspects of the brain's hemispheres consistent with the folklore of the symbolic left and right hands. Experiments suggest that the right brain has little sense of time or of cause-and-effect relationships; it can act without reason, is very intuitive, even holistic.

For most engineers, the left brain is much more familiar. It has a strong sense of time and of cause-and-effect relationships. Its rational basis finds expression in facts, numbers, and words. Our education, focusing on symbolic mathematics and verbal skills and performance, tends to strengthen the left hemisphere. Creativity, on the other hand, has strong links to the right hemisphere, where designs are generally created. Scientific and mathematical skills are necessary for analysis and testing, language skills are required for communicating the designs to others; all are a function of the brain's left hemisphere.

Relaxed Attention

It is certainly very desirable, even necessary, to integrate the abilities of your left and right brain. Relaxed attention assists this integration. The concept almost seems like an oxymoron: How can you be relaxed and pay attention at the same time? The relaxing, the letting go, refers not to the activity or the thoughts you are following, but rather to the distractions, the thoughts or activities you do not wish to follow or involve. A virtuoso performer, be she an athlete or pianist, often elicits the remark, "She makes it seem so easy, so effortless." This performer is able to use all her energy for the performance; none is wasted on nonessential tensions. Needlessly tensed muscles not only divert attention from the performance, but also consume more energy than they would in a relaxed state. Thus, being uptight or tense will interfere with your thought process.

Relaxed attention is physical, emotional, and mental, as we need the unimpeded flow of ideas between mental (seeing), emotional (imagining), and physical (sketching) states. We should try to see clearly, sketch freely, and imagine unboundedly. Eyestrain and tiredness affect not only visual thinking, but also mental processes. Relaxation techniques, for example, yoga and transcendental meditation, allow you to direct and maintain attention on a given task or purpose. Your mind does not wander. Forcing attention, while sometimes necessary, robs you of energy as your mind tries to seek escape to other thoughts, clearly denying you creative mental interactions.

The most menacing problem facing a creative person, whether a writer or a problem solver, is a blank piece of paper, as it tends to stifle imagination. There are some guidelines that you can follow to enhance your creative abilities. Encourage fluent and flexible ideation, meaning many (fluent) and varied (flexible) ideas. Do not edit them. Do not judge the value of your ideas. Your right brain needs to wander and not be edited by an overriding left hemisphere. Try to create as many ideas as possible—quantity matters. Your skills in sketching are a necessary asset, so the drawings will be fluid and unhesitating responses to your ideas. When the ideas are ephemeral and at times fast-moving, your sketching needs to keep pace, or else it can become a brake on your imagination. These are sketches for interacting with your mind, not for communicating ideas to others, and need only capture images for you.

Creativity allows us to be open to new ideas, new ways of viewing old products. It is this openness that allows good designers to change the original design concept as the process proceeds; there is not an ego investment in making something work when

a better alternative appears. This is not to say that persistence is
unnecessary; it certainly is, as noted earlier.

Design Process Overview

The design process starts with a human need that requires a
solution, a problem to solve. The problem may be narrowly defined, such as to refine a student desk lamp, or more loosely
characterized, such as to improve the water quality in a pond.
Before creating solutions, an engineer will research and investigate the topic. This research can be technical in nature, and it
may be nontechnical as well, to better understand the context
that created the problem. For instance, what factors contribute
to the pond's low water quality—development resulting in decreased water supply or increased nitrates from lawn fertilizer?
Were there problems with a previous version of a desk lamp, did
customers complain about its tippability, were there burns from
touching it? Technical matters are more readily determined, but
perhaps not as philosophically interesting.

In the case of the desk lamp, this may involve looking at other
types of desk lamps and learning about their characteristics, good
and bad. This leads to the specifications and clarifications area of
the problem statement; it is here that the output requirements
(specifications) are noted, such as provide X lumens of illumina-
tion 3 feet from the lamp, be adjustable in height between 15 and
30 inches, use fluorescent bulbs, be available in a variety of colors.
Constraints may be imposed that limit the variety of solutions
possible; perhaps the material must be molded plastic, and the
cost must not exceed $5. Most often the problem is presented to
the engineer by a customer along with specifications, what he or
she wants the design to provide.

At this point, the engineer more thoroughly understands the
problem technically through investigation and philosophically
through the design requirements. Now the creative side of the
designer is freed to brainstorm, to create several quick sketches
leading to different ways of solving the problem. This is perhaps
the most challenging part of the design process, as we often
seize upon one idea, judge it to be satisfactory, and cease to
think creatively, searching for the best or optimum design. A
corollary in mathematics to the optimum is finding whether a
point on a curve is a maximum (best, optimum) by taking the first
derivative and seeing if it changes slope. In the creative sphere,
other designs provide a similar validity for the optimum one,
designs to check against. In this process positive and negative
features of the designs are examined, and the best or optimum
design is determined.

Before the design is constructed, it needs to be analyzed. Will it
perform the required functions, meeting the design specifications?

The analysis of the mathematical model of the physical system is one of the important components of engineering design. The analysis of the model will include use of the constraints that have been developed in the problem definition, such as the forces acting on a structure, or an electric signal to be interpreted. The questions answered here include whether the device will meet the functional requirements desired. Perhaps an original goal is too demanding, such as a lightweight structure that supports too great a force, and the material constraint may have to be revisited. The analysis portion of the design process will feel most familiar to students, as much of the course work in engineering relates to engineering analysis, for example, circuit analysis, structural analysis, and thermal analysis.

Once the analysis is completed with any modifications made to the design as a result of the analysis, the construction begins. The process itself, shown in Figure 2.3, is nonlinear, as is typical of creativity and in contrast to the linearity and logic of mathematical analysis and scientific inquiry. Note the inner arrows on Figure 2.3; these indicate that at any time in the process, it is fine to go back and add information, perhaps adding specifications, eliminating a constraint. Invariably during the construction of the device, changes are made to the conceptual design. It is important to note these and recognize that such changes are fine—they are part of the creative process. Of course, you must make certain that the device solves the problem, and this is where the testing is done. The testing should be conducted in a scientifically correct fashion, so the data are valid and reliable.

At this point you know whether the device meets the problem statement specifications and satisfies the constraints. Were this a company, a management decision would be made to proceed with manufacturing or not. It may appear strange that a project could be canceled after a good deal of creative effort has been

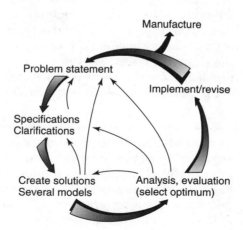

Figure 2.3
The design process.

put into it, but there are many possible reasons for this. The project's budget may have been exhausted, and the engineers transferred to another project; or it has taken too long to develop the product, and from a marketing perspective the release needs to be delayed; or a competitor using newer technology renders the design noncompetitive.

As a student, you may work with others on a project or by yourself. However, in industry you will seldom have the responsibility to conceive, design, and manufacture a product; it is simply too large an undertaking for one person. There are engineering specializations associated with each of these areas. In traditional organizations, research and development proved the feasibility of designs, and implicitly they are required by the company; design engineers created alternative designs and selected the optimum one subject to constraints and specifications. The design engineers also created the detailed drawings and specifications for manufacturing to implementation. Manufacturing engineers, in turn, modify the drawings in light of the machinery and personnel capability to produce the product, and sales engineers are involved in the marketing and sales of the completed product. This linear system creates manifold problems and delays, particularly between the design and manufacturing engineers.

Consider an everyday product, the desk or table you are working on. Imagine yourself as the design engineer who had to communicate to someone else how to make that which was visualized. Perhaps the table is 30 inches (in.) by 60 in. with a metal base and folding legs. Is the desk exactly 30 in.? Of course not. Any measurement has an error associated with it. What are satisfactory measurements and tolerances for this desk, consistent with the manufacturing equipment available? To answer requires experience and communication, perhaps 30 in. plus or minus $\frac{1}{16}$ in. is fine. The drawings will reflect this tolerance. However, a novice design engineer might specify 30 in. exactly or with a tolerance of 0.001 in. The manufacturing engineer then writes a memo, discussing the problems with the design and why it may need to be reconsidered, creating delay and adding expense to the design process. This is a very simple design. Imagine the problems that can occur in more sophisticated designs if there is not good communication between the various areas!

Concurrent Engineering

The traditional design process is being replaced by concurrent engineering design in which everyone involved in producing a product, from inception to shipping to sales, works as a team. It has long been recognized that changes occur during the design and manufacturing process; with concurrent engineering many of the changes occur earlier in the design process because of the increased communication between all parties at various stages

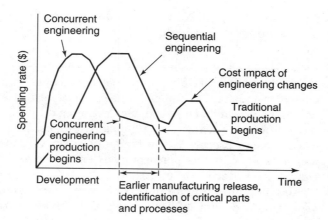

Figure 2.4
Comparison of traditional and concurrent engineering, spending rate versus time.

of production. Team-building meetings and discussions of the project's progress provide much more than traditional written communication. Thus, in the example of desk design and production, engineers in manufacturing communicate the machinery limitations to the design engineers before final drawings and specifications are created. Similarly, personnel in shipping can indicate savings achieved by the use of standard packaging if certain dimensions are incorporated. A major aerospace company found out that changes that cost $1 when made early in the development cycle of an aircraft rise to $10,000 during the manufacturing phase. During the production phase, many plans must be changed, resulting in costs associated for these changes as well as time lost from manufacturing.

Concurrent engineering reduces the time of the manufacturing cycle, which is increasingly important in the drive to get new products to market as quickly and reliably as possible. Figure 2.4 illustrates the spending and time profiles for traditional, sequential engineering and concurrent engineering. Note that the spending rate is highest early in the cycle for concurrent engineering, though the total cost (the area under the curve) is less than that in sequential manufacturing. A mind shift is required by managers when adopting concurrent engineering, not only in regard to costs, but also involving many people in the decision-making process and providing training time and meeting times for them, all of which costs money. The meetings and training are required to break down traditional, hierarchical methodologies used in manufacturing a product and to allow meaningful input, respecting the advice of people from all areas of design, manufacturing, and sales. The increased emphasis on teamwork is one of the newer characteristics of the modern engineering company. An engineer working in isolation is an idea whose time has passed. Teamwork allows the creativity of many people to inform the process, from design conceptualization to product fabrication. New ways of doing things that yield a better product are always sought.

Technical Decision Analysis

It may be comparatively easy to select the optimum design if one clearly stands out as the best choice on several fronts, but there will often be competing designs that offer different features and have different advantages and disadvantages. One characteristic of the design process is the trade-offs, cost versus material versus reliability versus performance. This contrasts with a single evaluation criterion, often cost.

An entertaining and informative way to introduce the concept of optimum design or best choice is by evaluating a consumer product, for instance, determining the best potato chip, soda, or chocolate bar. Not all the elements in the design process apply, but the important ones do. The problem statement is to determine the best potato chip. The investigation phase will probably be minimal, as you know the attributes of potato chips that appeal most to you. When you are brainstorming for ideas, the thoughts are about attributes of potato chips, things you like or dislike about them. A list often includes saltiness, crunchiness, thickness, size, fat content, calories, cost, color, and taste. There are no alternative solutions, but there will be various criteria that you think are important, perhaps saltiness, crunchiness, fat content, and cost. Explain why these criteria are the most important to you, and then devise a testing methodology that will reliably test a variety of potato chips for these criteria. The sodium content is listed on the packaging, so this value is readily obtained. You will have to decide if more or less salt is better, or perhaps there will be an optimum salt level that emerges from the taste test. Consider an attribute that is not so readily found, such as crunchiness; how might that be determined? Some have suggested repetitive testing by people chewing a chip, to find out if there is a concurrence on one brand as the crunchiest. Others have suggested using a decibel meter and breaking the chip at a fixed distance from the meter, the loudest breaking sound being correlated to crunch. Sometimes the criteria are too difficult to evaluate quantitatively with the equipment available; perhaps you do not have a decibel meter, and alternative testing means must be sought. During the evaluation phase, there may be feedback to previous elements of the design process. You may want to change the criteria for assessing the chip's quality in light of reliable testing methodologies.

At this point, assume that the tests have all been conducted in a scientifically reliable manner, and the results are in: Chip A is crunchiest, chip B costs the least, and chip C is the saltiest. How do you determine the best? The multicriteria decision analysis technique allows us to balance these in a rational fashion, arriving at a best or optimum product. The exercise of determining the best chip can be fun and provides insight into the human-made world, a world of trade-offs and choices, as the chip you like the best may not be the chip your best friend likes the best, but the process of determining the best in both cases fundamentally follows the design process.

Consider a recent situation you most likely found yourself in, selecting a university to attend. How might you decide between several in a rational fashion? The initial step, similar to definition of the characteristics of a potato chip, is to determine independent criteria used in making the decision. Not all criteria are equally important, so the relative importance of each criterion is decided upon, denoted by weighting factors. Table 2.1 lists some typical criteria and their weighting factors. To determine the criteria and the weights is, in general, not an easy task if skewing of the decision is to be avoided. Duplication of criteria may occur, such as reputation and quality of the faculty.

The following criteria were used in selecting the optimum university: Cost is a very important criterion and deserves a top weighting factor for most people. Nearness to home may or may not be important to you and was given a lower weighting factor. One of the keys to success at a university is how you relate to your fellow students; if you find a compatible group of peers, you will be much more likely to have a successful college career, again an important factor. The size of the university may not be as important to you. Whether the college or university comprises 2000 students or 20,000 students may be less a concern than whether it has the major you desire or enjoys the stellar reputation that many of its faculty and graduates have. The importance of the factors to you will give them different relative weights.

At this point, you can select several alternative schools and rate them with a score of 1 to 10 as to how well they perform in each of the six categories. The score is multiplied by the weighting factor, and the total for each school is determined, with the highest total score being the best school based on these criteria

Table 2.1 Criteria and weighting factors for selecting a college

Criterion	Weighting factor (1–10)
1. Cost	9
2. Class size	6
3. Compatible student body	7
4. Major desired	8
5. Reputation	7
6. Nearness to home	5

and scores. For this example, let the contrast be between a large public university (LPU) and a medium-size private university (MPU). Table 2.2 indicates the scoring that might occur. In this situation the tuition cost of the public university is as inexpensive as possible, except for perhaps community college for the first two years, thus yielding a high comparative score. The number of students enrolled is much larger, with larger class sizes, hence the higher rating for the private university. The student body characteristics at the private university were more appealing to this hypothetical engineering student. On the other hand, the LPU did have exactly the major desired (e.g., ceramic engineering), whereas the MPU did not (e.g., materials science). Again the reputations of the faculty and graduates from both institutions were very good, with the edge going to the LPU because of more research and publications. The MPU was nearer to home, a plus in this instance. You can see from the totals that the two universities are virtually tied in the rating, indicating the difficulty of making a decision. It is possible for bias to enter the scoring; however, there are ways to normalize the decision making, minimizing this effect.

This type of normalizing is referred to as *benchmarking,* which entails establishing what is the best for a given criterion, quantifying it, and then comparing a given product, process, or college to the benchmark. For instance, in considering four-year undergraduate schools, the cost at the LPU is $10,000 per year whereas at the MPU the cost is $20,000 annually. As noted earlier, the LPU has the lowest possible cost, hence its tuition becomes the benchmark value. To determine the rating of the universities, divide the benchmark value by the university's cost, and multiply by 10 to convert the number to a value between 1 and 10. For instance,

$$\text{LPU cost rating} = \frac{\$10,000}{\$10,000} = 1.0 \qquad (1.0)(10) = 10$$

Table 2.2 Weighted comparison of two universities

Criterion	Weight (1–10)	Private university	Public university	Private university	Public university
Cost	9	5	10	45	90
Size	6	10	5	60	30
Student body	7	8	6	56	42
Major	8	7	9	56	72
Reputation	7	7	8	49	56
Closeness to home	5	8	5	50	25
			Total	316	315

$$\text{MPU cost rating } = \frac{\$10,000}{\$20,000} = 0.5 \qquad (0.5)(10) = 5$$

The average class size in the MPU is 25 students, again viewed as the lowest possible and hence the benchmark value; and the LPU has an average class size of 50.

$$\text{LPU class size rating } = \frac{25}{50} = 0.5 \qquad (0.5)(10) = 5$$

$$\text{MPU class size rating } = \frac{25}{25} = 1.0 \qquad (1.0)(10) = 10$$

The analysis is very adaptable to spreadsheets so that adding universities is not a chore. Determining benchmark values requires research, as the quality of the analysis is a function of the quality of the data used.

Ecological Design

A trend in engineering and other design professions, such as architecture, is to seek designs that are compatible with the natural environment in their development, operation, and disposal. This springs from the growing recognition that the affluence we currently enjoy is, in part, borrowed from the future, because we have been depleting natural resources, minerals, and energy at an irreplaceable rate and in the process are creating an unsustainable world. Some consider that we are living in two worlds. The first is the natural world evolving over 4 billion years. The second is the human-made world of roads and cities, farms, and artifacts designed over the last few millennia. The condition that threatens both worlds—unsustainability—results from a lack of integration between them.

Ecological design is simply the effective adaptation to and integration with nature's processes. It tests solutions with a careful accounting of their full environmental impacts. There are now sewage treatment plants that use constructed marshes to simultaneously purify water, reclaim nutrients, and provide habitat; as well as agricultural systems that mimic natural ecosystems and merge with their surrounding landscapes. New types of industrial systems have the waste stream from one process designed to be useful input to the next, thus minimizing pollution.

Engineers with an ecological focus observed that the traditional model of industrial activity—in which individual manufacturing processes take in raw materials and generate products to be sold, plus waste to be disposed of—should be transformed to a more integrated model—an industrial ecosystem. In such a system, the consumption of energy and materials is optimized, waste generation is minimized, and effluents of one process—whether they are spent catalysts from petroleum refining, fly and bottom ash from electric power generation, or discarded plastic containers from consumer products—serve as raw materials for another process.

Engineering traditionally concerns itself with safety and efficiency; ecological design asks that we do more. We have already made dramatic progress in many areas by substituting design intelligence for the extravagant use of energy and materials. Computing power that 50 years ago would fill a house full of vacuum tubes and wires can now be held in the palm of your hand. The old steel mills whose blast furnaces, slag heaps, and towering smokestacks dominated the industrial landscape have been replaced with efficient scaled-down facilities and processes. Many products and processes have been miniaturized, dramatically reducing the flow of energy and materials required to fabricate and operate them. For instance, there is manufacturing using molecular nanotechnology. If we think about products at their most fundamental level as being made from atoms, then the properties of these products depend on how the atoms are arranged. If atoms of coal are rearranged, diamonds may be created. Similarly, by rearranging the atoms of sand with a few trace elements added, computer chips are made. Today's manufacturing methods are very crude when seen from a molecular view—casting, grinding, milling, and even lithography move atoms in huge statistical groups. Molecular nanotechnology, or manufacturing, allows positioning of every atom in its correct place, yielding reduced costs in terms of materials and energy.

Of course, society plays a pivotal role in creating the climate for change, making it economically feasible to use intelligence of ecological design. For instance, in Germany, manufacturers are now required by law to either take back and recycle old packaging or pay a steep tax. This has transformed the German packaging industry. Questions now include: How can durability and reuse be designed into the packaging? How can easy disassembly of packaging components facilitate recycling? These questions have triggered extraordinary innovations in reusable or recyclable packaging with corresponding environmental benefits, including decreased waste and use of virgin materials. Traditional design does not ask these questions, instead optimizing with respect to cost or convenience with minimal or no environmental considerations.

In a sense, evolution is nature's ongoing design process. The wonderful thing about this process is that it is happening continuously throughout the entire biosphere. A typical organism has undergone at least a million years of intensive research and development. A few years ago, two Norwegian researchers set out to determine the bacterial diversity of a small amount of beech forest soil and of shallow coastal sediment. They found well over 4000 species in each sample, which more than equaled the

A petroleum engineer checking the pressure differential from an oil rig. (*Courtesy of Texaco, Inc.*)

number listed in the standard catalog of bacterial diversity. Even more remarkably, the species present in the two samples were almost completely distinct—nature's design process with different constraints yielded different solutions.

In the sustainable design of products and processes with an eye toward ecological awareness, three strategies are employed to address the environmental effects: conservation, regeneration, and stewardship. Conservation slows the rate at which things are getting worse by allowing scarce resources to be stretched further. Typical measures are recycling, adding insulation, and designing fuel-efficient cars. Regeneration is the expansion of the natural capital by the active restoration of degraded ecosystems and communities. Stewardship is a practical quality of care in relation to other living creatures and to the landscape. Stewardship requires continual reinvestment, observation, and design innovation. An example might be how to control erosion on a steep hillside. A conventional civil engineering design could use a thick concrete retaining wall to hold the earth. In ecological design, natural processes are used to accomplish the same goal, such as seeding the hill with willow branches which sprout and develop articulated roots, keeping the soil in place. However, the willows must be tended until the roots develop; a knowledge of tree characteristics is required to pick willows with deep root systems and not another species with shallow root systems. Notice that the ecological solution uses very little energy and matter, but does require stewardship, while the traditional method uses much more energy and matter, but requires very little stewardship.

We, as a society, have expected design professionals to bend an inert world into shape. An alternative is to try to catalyze the self-designing potentialities of nature, allowing useful properties to emerge rather than deliberately imposing them.

Innovation and Quality

Engineers have been very inventive in creating the human-made world. One of the criticisms of companies and indirectly of engineers in the United States in the recent past is that many devices were created here, but eventually the manufacture and subsequent innovations to the inventions occurred in other countries. A tremendous insight is gained into the original design as the manufacturing and construction process occurs, modifying the final design. As the object is used, problems may arise with its operation, and features may be desired to enhance the operation. The people who manufacture the product will gain the knowledge necessary to make these improvements and with this the ability to create new and better products. The chain is important, as the concurrent engineering model indicates. To have design without feedback from manufacturing is shortsighted, and the design will be short-lived. Engineers did not create the corporate strategy of the 1980s to shift manufacturing to other countries or outsourcing to other companies, but their employment opportunities were certainly affected. Information technologies permit the linking of engineers in Europe, Asia, and North America. They may all be working on the same product, perhaps a common automobile design for all countries. Drawings and messages are sent electronically on secure Intranet connections.

One of the reasons for upgrading designs, or innovating, is that customers use the product in unanticipated ways, expecting outcomes not initially conceptualized. A feature of good designs is their robustness, in the design sense meaning the ability to perform well under unusual circumstances. Robustness is particularly needed in complex systems, such as those found in nuclear power plants where there must be no possibility of unanticipated failures resulting in an accident. This is made additionally difficult because the science of complex systems indicates that these systems undergo reconfigurations and realignments. Chaos theory tells us that even if we have an exact and deterministic model of a system that is completely closed to outside influences, its behavior will not be predictable beyond a certain time scale. In the case of systems that are subject to outside influences, such as power fluctuations of the electrical grid or air and water quality changes, the time scale becomes shorter. An analysis of the safe management of complex systems points to one of the features of ecological design, stewardship, as well as to features of concurrent engineering. In the safe operation of complex engineering systems, there is a hierarchy, a chain of command, that addresses the routine matters and prevents errors of omission. But this in itself would not prevent complex systems from failing. Underlying the hierarchical chain of command, or set of rules, is empowerment of employees to question procedures and challenge beliefs, knowing that some ambiguity is inherent in complex systems and that it is necessary to keep people alert and involved. The involvement requires communication between a variety of fields, as the work on an electric generator may have ramifications for the control of the fuel rods. Thus, in addition to the hierarchical system, information technology allows a parallel structure of communication and organization adaptive in responding to changing circumstances. The information exchange allows an adaptive response, which parallels aspects of the ecological design example where stewardship of the willow trees is ongoing.

Safety and quality are issues of great concern to all engineers. Everything fails at some point, living organisms die, physical devices cease to function. The goal of engineers is to ensure that the failure is safe, a pump does not explode, an electric motor does not burst into flames. Whether the device is of high quality, often associated with a long lifetime, or low quality, its failure should be safe. Robust design is important, as the operating conditions may not be those anticipated. Quality is a corporate decision based, in part, on market conditions and cost. A company must be able to design products with a low enough cost to be competitive, or it may soon find itself out of business. For instance, seat belts were not always required in automobiles; they were an extra-cost option. Manufacturers did not believe they could add these as a standard feature and remain competitive. Government stepped in and passed a law requiring them to do so, and thus the challenge was changed to that of creating economical seat belt systems to meet or exceed the regulations.

Making a product safe—protection from electric shock, sharp edges, moving parts—is not the only consideration. The product should be comfortable to use, so ergonomics becomes important. Ergonomic analysis requires an understanding of human physiology to coordinate products to our physical abilities, so we are not in conflict with them. This can also extend to the work environment; if it is very noisy, hot, and humid, people will not be as alert as in quieter, more temperate environments.

Product Life Cycle

After market analysis indicates there is need for a new product and it is introduced to the market, most products go through a cycle of sales such as that in Figure 2.5. The initial cost of a device is usually much higher than its final price after being on the market for several years. For instance, when handheld calculators were introduced, they cost more than $200; now many with those same features can be purchased for less than $10. During the growth stage, customer feedback prompts design innovations, more devices are manufactured, the unit cost to produce them decreases, and more people purchase them. The maturity stage occurs as the price stabilizes, the product has features that most people desire, and sales increase. Eventually, many people who

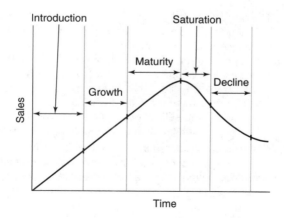

Figure 2.5
Sales projection pattern for a product.

desire the product have purchased it, and the product cycle enters the saturation stage. At this point in the product life cycle, there are more devices than can be sold, resulting in price decreases and, perhaps, the product's being taken off the market. The life cycle can be a matter of months, years, or many years, depending on the product. Engineers face the challenge of creating new products to replace the ones with declining sales, maintaining the company's profitability.

References

1. Arnheim, Rudolf. *Visual Thinking.* Berkeley: University of California Press, 1969.
2. Edwards, Betty. *Drawing on the Right Side of the Brain.* Los Angeles: J. P. Tarcher, 1979.
3. Eide, Arvid (ed.). *Introduction to Engineering Design.* New York: McGraw-Hill, 1997.
4. Kamm, Lawrence J. *Real-World Engineering.* New York: IEEE Press, 1991.
5. Koestler, Arthur. *The Act of Creation.* New York: Macmillan, 1969.
6. Koen, Billy Vaughn. *Definition of the Engineering Method.* Washington, DC: American Society for Engineering Education, 1985.
7. Lumsdaine, Edward; and Monika Lumsdaine. *Creative Problem Solving: Thinking Skills in a Changing World,* 3rd ed. New York: McGraw-Hill, 1994.
8. McKim, Robert. *Visual Thinking.* Belmont, CA: Wadsworth, 1980.
9. Minsky, Marvin. *The Society of Mind.* New York: Simon & Schuster, 1986.
10. Van Der Ryn, Sim; and Stuart Cowen. *Ecological Design.* Washington, DC: Island Press, 1995.

Problems

2.1. The purpose of this problem is to demonstrate and stimulate the need for fluency in sketching when expressing yourself graphically during the conceptual design process. Take a large piece of drawing paper and draw 30 freehand circles approximately 1 in. in diameter and 2 in. apart. Now in a timed exercise totaling only 5 minutes (min), sketch details on each circle that define it. Examples would be a flower blossom or a baseball. Could you complete all 30 circles in 5 min? Did you find yourself repeating patterns? Were you sufficiently fluent with your sketching that your ideas could be actuated?

2.2. There is a popular game called Pictionary in which you must at times sketch concepts. Make an abstract sketch of each of the following: animal, tree, shut, penetrate, collapse, turbulent, sharp, decayed, lively thrust. The idea is to communicate not to others but to yourself.

2.3. Very often we can form mental images of certain physical objects, but the clarity of the images fades because the thought is involved with more sensory perceptions. Note the clarity of your mental image for each of the following, from clear to nonexistent:

(a) Your mother's face

(b) A dandelion

(c) Your kitchen at home

(d) A sunset

(e) Children laughing

(f) The sound of rain on a metal roof

(g) A pinprick

(h) An itch

(i) Kicking a ball

(j) The taste of pizza

(k) The smell of gasoline

(l) Hunger

(m) Well-being

2.4. Not only is it useful to limber up your mental imagery as indicated in Problem 2.3, but also it is important to control images. In the following situations, you will find some you can control well, others not well, and still others not at all. Check your ability with the following:

(a) The concentric circles formed and expanding outward caused by a pebble dropped in the still water of a pond

(b) A flower blossom very slowly opening and blooming

(c) This text flying away into the sky and disappearing

(d) The chair you are sitting in coming alive and carrying you around the room

2.5. Preconceptions often limit our ability to see what is, what exists. Check your ability by

(a) Describing the telephone in your room, including the location of the numbers and letters

(b) Describing the dashboard of a car

(c) Describing the cover of this book

2.6. Select a consumer product and determine the best or optimum brand. Include the characteristics you believe are important and why you believe they are important. Select four characteristics for use in testing in a scientifically correct manner and the weighting factors you assigned each characteristic.

2.7. Write the specifications, including drawings or sketches if necessary, to construct the door entering your bedroom or dormitory room.

2.8. Develop the criteria and weighting factors for changing majors, buying a car, and living on or off campus. For hypothetical situations with the above, perform a multicriteria decision analysis to support the best choice.

2.9. Describe relaxed attention.

2.10. Consider a consumer product you purchased or examined in a store. How might the packaging be redesigned for reuse? For recycling? How might the product have different design features so that it could be recycled into component parts?

2.11. Investigate the use of marshes for waste purification. Consider the amount of land that is required as part of your investigation.

2.12. Discuss design intelligence and its effect on people's use of a device. If possible, examine a product from 30 or more years ago and the current version of that product. How is design intelligence manifested?

2.13. Innovation requires feedback from end users. Examine a consumer product that you are familiar with, and look for improvements that might be incorporated into a new model.

2.14. What is the difference between product safety and product quality?

2.15. Examine several products you are familiar with that have different manufacturers or models. What features indicate higher quality versus lower quality? After you examine several products in this way, are there any generalizations you can make about high-quality versus low-quality attributes?

2.16. It is often possible to improve the ergonomic design of devices we use. Pick one such device and find ways to improve it ergonomically.

2.17. Cut two pieces of paper or light cardboard into six geometric shapes with straight sides such that the sides are equal in length to one another or multiples of one another. Two people sit on opposite sides of a table with a cardboard divider between them, each with one set of shapes that cannot be seen by the other. One of you will act as the design engineer, the other as the manufacturing engineer. The design engineer will create a design, or pattern, and must communicate this to the manufacturing engineer without the aid of sketches or hand signals. The manufacturing engineer fabricates the design based on this information. Note the precision in communication necessary for a simple design to be replicated. Switch roles and repeat the process.

3

Design Documentation

The designs that engineers create require documentation, as the designs may lead to new patents; reports have to be written describing the what, when, where, why, and how of the project; and a record must be kept for further work on this or related projects. There are several ways that documentation occurs, principally through use of the design notebook or journal and the final report. In addition, the design portfolio is introduced as an aid to gathering the information necessary for the report; although a portfolio is not used in industry, it is valuable in school.

Written Communication

As an engineer, you will want to persuade people to follow your recommendations, act on your requests. In reviewing your analysis of a bridge or circuit design, few managers will question the wisdom of your calculations (although supervisors or colleagues will check them for accuracy), but they may question nonmathematical areas of judgment used in the assumptions framing the calculation. If a certain linkage is important in the design, though not the only way to accomplish a task, you should make a persuasive case for it, or it may not be accepted. Embedded within the design process is the selection of the optimum solution among several possibilities, and it is important to articulate well your justification for this solution.

In business and industry, a competing view may be asserted by another department asking for funding, recognizing that not all projects can be funded. The marketing department may want assistance to expand sales; manufacturing may request funds for new equipment or, in the case of a design, argue against the design because it requires new labor skills. The reports that you write describing your project, as well as the memos and letters that you generate, should persuade others that you are correct, your reasoning is sound, and your conclusions are convincing. Of course, the work you generate must be good, it must stand on its own merits; and you will ensure this standing if those merits are effectively communicated.

It is not redundant to ask yourself before writing whether you have a good idea of precisely what you want to communicate. Frequently, this will relate to presenting information or persuading people to act or think in a certain way. Questions to ask yourself at this stage include: Why do they want to read this document? Does the technical level of detail intrigue or bore them? Is their initial attitude likely to be positive, negative, or neutral? Will the document give them the information they desire? Answers to these questions will help you write more effective term papers, letters, and engineering reports.

Get to the point. Readers appreciate going from the general to the specific, from conclusions to details. For instance, if you perform a series of tests on a piece of equipment to determine its desirability in a system, your supervisor will want to know what you found out and your recommendation. The supporting detail of the testing procedure is important and will be further analyzed by others, but as the recommendation moves upward in management, your conclusion is the focus. Please note, this assumes that the testing procedure and the analysis are sound and accurate; if they are not, then no amount of clever language will suffice. Engineering competence is essential. Also, distinguish between fact and opinion, supporting the latter with facts. For instance, stating that the Hewlett-Packard graphing calculator is the best calculator is an opinion; how much it costs, consumer testing reports, and various features are facts that support your contention.

Express yourself clearly, getting to the point; don't allow misinterpretation to occur because a sentence has more than one meaning. Ambiguity, vagueness, and lack of coherence or directness may leave a statement open to several interpretations and not bolster your point of view. You are persuading the reader to your viewpoint, not a point of view unsupported by and inconsistent with the rest of the document. Here is an example of ambiguity: "Before accepting materials from new subcontractors, they should meet our requirements." Who is *they*—the materials or the subcontractors? A better way to write this is: "Before we accept them, the materials from the new subcontractors should meet our requirements."

How far behind are you on the project? Several weeks, or three weeks? The latter is specific, not vague. Specificity helps in problem solving; the person who reads the phrase *several weeks* will have to ask you, How many weeks is several weeks? Being direct and using active verbs invite the reader to grasp the points quickly. The passive voice is never as engaging as the active voice. Consider the following: "Water molecules are vaporized by heated wires." Compare this to: "Heated wires vaporize water molecules." Or compare: "The cylinder is closed at both ends by thin metal caps" to "Thin metal caps close the cylinder at both ends." Active verbs almost always provide direct and specific communication.

Your sentences need to be coherent and to form coherent paragraphs, which, in turn, form a coherent report. You achieve paragraph coherence by having any given sentence relate to the one before it and the one following it. The paragraph opens with a main point, and the sentences supporting that point link together, reinforcing that point.

Aesthetics matter. The appearance of what you write on the page influences the reader. Leave ample margins—the standard is 1 in. all around. A "ragged" right-hand margin is more readable than one that is right-hand-justified. Pick a typeface that is readable and familiar. There are two general formats for typefaces: *serif* and *sans serif*. The serif typefaces, like the one used for this text, have little edges and curlicues on the letters and are used in books, newspapers, and magazines; hence readers are comfortable with them. Sans serif typefaces, like that used in this book for examples, do not have these edges, are less readable, but are very effective for titles and headings in which the font size is larger. A 12-point typeface is standard; use other sizes for special effects. White space is important, isolating and emphasizing important data and information and providing a visual resting space between blocks of information.

Design Notebook or Journal

Engineers should, and in many instances must, maintain a notebook that is a record of their daily work. This becomes an official document in the eyes of courts should patents arise from their work or litigation ensue about a project an engineer worked on. When engineers are working on nondesign matters, the notebook entry may be a summary of what was accomplished during a day, including a log of phone calls with clients. It serves as a means for refreshing one's memory, should questions arise about when something was done, or not done. Engineers use the notebook for any written design work and when working on a computer. Since many engineers design using both media, there are software programs that log the work. The notebook has sewn pages, so it is not possible to insert a page and it is apparent when a page is removed; thus the record cannot be tampered with, making it creditable evidence in legal matters. In Chapter 5, conversations with two engineers highlight the importance of a design notebook.

Design Portfolio

The design portfolio is a way of recording the information you gather, the creative thoughts you have, the testing and analysis you perform, and the construction you make as part of the design process. It also serves as a ready guide to make sure you included all elements in the design process and did not skip any key steps. Note that a portfolio is a device to assist students to document the design process; design portfolios are not used by practicing engineers. The elements in the portfolio are as follows:

- *Problem statement.* This describes the problem you wish to solve so that someone else can understand it, including what it will accomplish and any specifications or constraints.
- *Investigation.* List some questions that need resolution before proceeding with the solution, and research answers to these questions and how others may have addressed the problem in the past.
- *Brainstorming.* Once you have developed a knowledge base about the problem, use your creativity to come up with several possible ideas about or solutions to the problem.
- *Alternative solutions.* The optimum solution is the best of several possible solutions; present alternative solutions to the problem.
- *Optimum solution.* Select the best solution, and explain why you believe this to be the optimum one.
- *Construction.* Explain how you will construct your device and changes to the design that occurred during the construction phase.
- *Analysis and testing.* Did the device solve the problem? How did you test it, and what were the results?
- *Final evaluation.* Is this the best design? If you were to start again, what would you do differently?
- *Oral communication.* Plan a presentation to the class; include how you make the presentation and the overall content of the presentation.

Students were given the challenge to design and construct a device that would be of assistance in the kitchen at home to solve a problem. One student found that opening jars was often a problem for his mother and decided to create a jar top opener. His efforts are documented in the following design portfolio with the assessment of it following.

INTRODUCTION TO ENGINEERING

DESIGN PORTFOLIO

Name   Ciro Poccia

Section 3A

Problem Statement

In your own words, describe the problem clearly so someone else can understand it. What will the solution accomplish? Are there imposed specifications and limitations?

Often jar tops are difficult to open with the bare hand, particularly for older people or people with arthritis or physical impediments. The jar top opener needs to increase the torque beyond that of the bare hand. The handle should be easy to hold, and the opener should securely grip the jar cap and not slide. The device needs to be inexpensive, costing less than $5 in materials for the prototype. Since it will be used in a kitchen, it must fit in a kitchen cabinet drawer.

Investigation

What are some questions that must be answered to solve the problem? List at least three.

What will increase the torque in turning the cap?

How or in what way will the device grab the cap?

How will the handle be incorporated into the actual device?

Resources

Make a list of resources that you have used to obtain information about the problem to answer questions you listed above. These may include people, written material, or electronic media.

I questioned my mother about the jar openers and features she would like.

I consulted an elementary statics book on forces and torque.

Information

What information have you gathered from the resources noted above?

1. *Information from people.*   My mother had the idea of using a rubber glove to increase the grip around the cap. She mentioned wrapping an elastic band around the cap to improve the grip. She wanted a device that was easy to use, did not require instructions, and worked on all sizes of containers. Further investigation of the containers found that cap size varied from 1 to 4 in. in diameter.

2. *Information from written material.*   From a statics book, the concept of having a longer handle for more torque was derived. Torque equals force (from the hand) times distance to the center of the cap.

3. *Information from electronic media.*

Brainstorming for Ideas

Sketching is a great way to generate ideas. Use the space below to draw or sketch as many ideas as you can think of. At this point, do not eliminate anything that may have possibilities. You may want to add additional sketches, perhaps using graph paper.

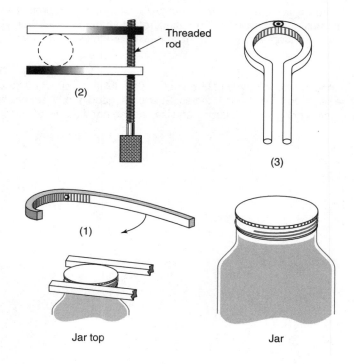

Alternative Solutions—Describe Your Best Ideas

Describe three of your most workable solutions to the problem. Remember to consider the specifications and limitations. In your description, indicate what you consider to be each solution's strengths and weaknesses.

Solution 1

The socket wrench idea has a handle that looks like an upside-down cane. The curved part will have a rubber lining and tighten around the cap. The tie rod extending from the cane will be mounted on a hinge with elastic bands for tension. This rod sets into a cap groove and applies force when the handle is turned. Strengths: unique and different idea. Weaknesses: hard curve to design, limited cap size. See sketch 1, page 53.

Solution 2

The threaded rod idea is basically like a monkey wrench with the top half of the jaw attached, yet not able to move with the tie-rod rotation. The bottom jaw will move with rod rotation to tighten against the cap. Then turning the handle provides the torque to turn the cap. Strengths: variable cap size, good grip, good torque. Weaknesses: may be awkward to use. See sketch 2, page 53.

Solution 3

The plier idea is to shape and build a type of large nutcracker. It will be attached by a top hinge, and when closed, the other end will be the handles. Strengths: good grip. Weaknesses: limited cap size, torque characteristics not good with large cap sizes as the handles will be difficult to grip. See sketch 3, page 53.

Selecting Your Best Solution

Describe your best solution, and indicate below why you selected this solution.

The best solution was the threaded rod that also functioned as a handle. Two pieces are mounted perpendicular to the rod. The one farther from the handle is attached yet does not move with rod rotation. The second moves up and down the rod when the rod is rotated. Both pieces are mounted to the rod from the top, so the cap is placed underneath. This solution allows a variety of cap sizes, the force gripping the cap can be increased as necessary, and it should be relatively easy and inexpensive to construct.

Describe how you are going to construct the solution to the problem.

A drawing will be made for each piece of the wooden jaws and shaped to conform to a variety of cap sizes. The wooden jaws, made from wood 1 in. by 6 in., will be cut and then shaped with a jigsaw. The nonmoving jaw farther from the handle will have an eyebolt placed in it. The threaded rod will slide through this. The tip of the rod will be drilled and a cotter pin used to prevent the rod from sliding out. The second jaw, the movable one, will have a nut glued to it and the rod threaded through it. A closet pole will be cut, drilled, threaded, and glued onto the rod to serve as a handle. A rubber glove will be cut and glued to the inside surface of the jaws to enhance the grip on the cap. The figure below illustrates the jar opener.

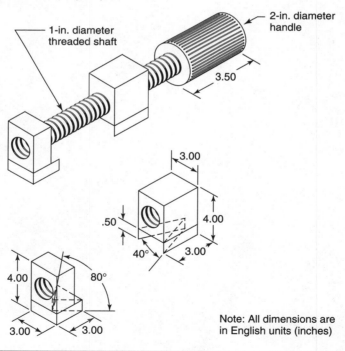

Note: All dimensions are in English units (inches)

Construct Your Solution

Construct your solution to the problem based on the specifications and limitations. Describe any modifications to the design you made as a result of the construction process. How did these alter the solution?

The original shape of the pieces was going to be a V, but using different-shaped caps proved that a _/ shape would be best, as the V had too dramatic a slope which did not allow the jaws to grip the cap well. Once cut, it was difficult to find an eyebolt to fit with the $\frac{3}{8}$-in. thread rod and perhaps would not be strong enough for the torque it would be subjected to. The wall piece to a deadbolt lock system was used, and this fit the rod freely and was very strong.

Next the nut being glued to the wood definitely did not seem strong enough, so the idea of two blocks of wood on either side of the nut seemed wise. Two more nuts were added for safety. The blocks were glued and squeezed together while the adhesive dried.

The handle was cut and drilled with rough edges filed. It was threaded on and glued for secure operation. The drilling of the threaded rod was difficult, yet it was accomplished. A washer was added before the cotter pin just as a precaution. The final product gripped the cap well and had plenty of torque.

Analysis and Testing

Describe how you will test the design to determine if it works. Plan to conduct the test more than once to ensure that the results are repeatable and not just due to luck.

The test was simply done—jars of different sizes were readily opened with a slight tug on the handle. The opener satisfied the criterion of working on differing sizes. The only disappointment in the testing was the time it took to adjust the jaws when positioned for a large-diameter cap and then repositioned for a small-diameter cap. The caps were retightened as much as possible by hand, and then the opener was used again. A slight effort was needed in all cases. I did not have a spring scale which I could attach to the handle to measure the force necessary to open the cap. My mother indicated that the opener required very little effort, although it was a little awkward to use in securing the jaws to the cap—two hands were required.

Describe the results of your tests below. Use graphs and charts as appropriate supplements.

Four different jar top diameters were used: 1.5, 2.5, 3.0, and 4.0 in. The larger the cap diameter, the greater the force necessary to open it. The large-diameter jars were the ones my mother had the greatest difficulty opening.

Final Evaluation

Did your solution solve the problem within the specifications and constraints? Explain.

Yes, the solution solved the problem, it increased torque over that of the bare hand, the design was inexpensive, and all the material was made from scrap found in my basement. I estimate the cost if purchased at $3.50, mostly for the threaded rod. The opener fit in the kitchen cabinet drawer, barely, and perhaps could be shortened for future models.

If it did not solve the problem, explain what you think went wrong. Remember, learning what does not work is often as important as knowing what does.

What do you like about your solution?

I liked that the opener worked well on a large variety of jar sizes and that not a lot of effort was required to open the jar. Large-diameter jars are often difficult to open by hand, and the device will be a great assistance to my mom.

What do you dislike?

Turning the handle took time if a great deal of adjustment needed to be made. Also, it was difficult to determine how much to tighten the jaws. They were tightened until firm resistance was felt, but that is imprecise.

If you had to solve the problem again, would you use the same solution? Why or why not?

I liked elements of the solution, but the overall length needs to be decreased. Since the force needed to open the jars was small, shortening the handle will allow it to fit more readily into kitchen drawers and still provide sufficient torque with little effort. The threading takes too long. Perhaps a sliding jaw such as used on furniture clamps can be used or threads with a larger pitch, so the linear movement is greater for each rotation of the handle.

Communicating to the Class

If you are communicating your solution to the class, what media will you use?

It will be an oral presentation and demonstration of the opener. I will bring a jar and show how the device works. There will be sketches on transparencies to show on the overhead projector, so I can point out the key elements in the design.

Outline the contents of your presentation below.

Problem statement and constraints, specifications.

Different ideas from brainstorming.

Why I selected the final design.

Problems encountered in constructing the opener, including use of materials and modifications to original plan.

Testing and evaluation of the opener.

Demonstration of the opener on a jar.

Assessment

Design portfolio assessment examines the process and the final product. An important goal in engineering design at the introductory level is to understand the complexity of the process. The end product is important, but it is unrealistic to expect initial designs to be unique, creative, analyzed well, and fabricated perfectly with a focus and judgment made on the end product. Designing is creative and interactive; the process elements support one another, such as the construction phase always lends additional understanding and usually modifications are made to the device. The portfolio documents the process. The assessment evaluates the process, denoted in the portfolio, and the device itself. The following assessment is for the jar opener portfolio and constructed device.

DESIGN ASSESSMENT RUBRICS

The Design Process

A. Identified problem criteria, constraints, and specifications.   0   1   2   (3)
Good understanding of problem, constraints, and specifications. Problem to solve is explained well.

B. Gathered background information from a variety of sources.   0   1   (2)   3
Did not examine devices already on the market (e.g., look in kitchen supply catalogs, visit store).

C. Suggested several alternative solutions.   0   1   2   (3)
Three alternative solutions that could work were suggested.

D. Evaluated ideas against design criteria and made improvements.   0   1   (2)   3
Did not make improvements on the original design based on testing; did make alterations based on construction process.

E. Justified the chosen solution.   0   1   2   (3)
Compared the advantages of alternative solutions, selected optimum design.

The Design Solution

A. Provided an accurate drawing with basic details and dimensions.   0   1   2   (3)
Elementary computer-assisted design drawing with overall dimensions provided.

B. Constructed the model and used materials appropriately.   0   1   2   (3)
Model well constructed, good use of hand tools and materials.

C. The solution worked. It fulfilled the design criteria.   0   1   2   (3)
The design worked well, it met all the design criteria.

D. Originality and creativity of the design.   0   1   (2)   3
The design was a bit difficult to use, threading of rod for differing jar tops was time-consuming.

Testing

A. Used knowledge gained from testing to inform design.   0   1   (2)   3
Good testing, but did not change design in light of testing information.

Work Habits

A. Completed assigned task in a timely fashion.   0   1   2   ③
Project and design portfolio completed on time.

Communication and Presentation

A. Demonstrated understanding of key ideas orally and/or in writing.   0   1   2   ③
Understood problems of torque and friction, and presented them well. Engaging and well-prepared presentation.

B. Report neatly written with good grammar.   0   1   2   ③
Portfolio completed. All topics fully discussed. Grammar and spelling fine.

Scoring guide: 0 = No response or unacceptable response
1 = Acceptable response
2 = Good response
3 = Excellent response

Score: 35
Total possible points: 39

Design Report

Engineering reports have a certain format, or specifications, so readers will know what sections to expect and where to find information they are seeking. The reports may seem repetitive at times, but in reading reports, engineers skip around and do not necessarily read the report from beginning to end in linear fashion. This is particularly so as the report moves up in management; fewer people will read the report in its entirety, yet you want them to encounter the key facts and conclusions regardless of their reading pattern. The standard sections of an engineering report are

Transmittal letter
Covers and label
Title page
Table of contents
List of figures
Executive summary, abstract
Introduction
Body of the report
Appendixes (including references, charts, tables)

The transmittal letter is not something you will be concerned with as an engineering student writing design reports, but it should accompany any report you create as part of an internship, for instance. The letter explains the context for the report and why the report was necessary, and it very briefly states that a report is enclosed, a brief overview of the report, and whom to contact if there are questions. Figure 3.1 illustrates a transmittal letter.

Reports that are greater than about 10 pages should be bound with a label on the cover with the report title, your name, company, and date. Ideally the cover will allow the report to open flat, so spiral-bound covers are ideal. At the opposite end are the clear and colored plastic slip covers which do not allow the report to lie flat and often fall apart. For engineering course reports, bookstores will often have a variety of covers; perhaps the most useful are those for which you punch holes in the report paper and insert the pages in self-contained brads inside the cover. In this circumstance, increase the left-hand margin $\frac{1}{2}$ in. for the binding.

The title page of the report will contain the full title of the report; the name, title, and organization to which the report is submitted; your name, title, and organization; the report submission date; and other required information (perhaps a contract number). Figure 3.2 illustrates a cover page. Companies will undoubtedly have a style they use, but it will have the same elements as discussed in this section. Engineering design reports done as part of your course work can follow these guidelines, unless modified by the instructor.

300 Northern Boulevard
Great Neck, New York 11021-5066

May 7, 1998

Mr. James Harwick, Chief Engineer
ABC Technical Services
P.O. Box 3311
Pittsfield, MA 01262

Dear Mr. Harwick:

Enclosed is the report *Transient Thermal Response of HX Heat Exchangers* that Trident Engineering Associates performed per our contract agreement with you dated September 11, 1997.

The report focuses on the thermal response of HX heat exchangers should there be a failure to the circulating water pump in the emergency diesel generator cooling system at the Stony Ridge power plant. The report indicates the response time needed in several scenarios to continue running the diesel engines.

My colleagues and I hope that the report satisfies your needs and provides further insight into conditions necessary for safe engine operation. Should you have any questions or concerns, please feel free to contact me.

Sincerely,

Joan Daley, P.E.
Manager, Trident Engineering Associates

Encl: Report on Transient Thermal Response of HX Heat Exchangers

Figure 3.1

Report
on

TRANSIENT THERMAL RESPONSE OF HX HEAT EXCHANGERS

Submitted
to
Mr. James Harwick, Chief Engineer
ABC Technical Services

Prepared by
Trident Engineering Associates, Inc.

May 7, 1998

Figure 3.2

64

Technical reports may have a descriptive abstract that provides an overview of the purpose and content of the report or an executive summary, which is an expanded abstract, that summarizes the key facts and conclusions in the report. Typically, the executive abstracts are one to two pages long and may contain bulleted information that highlights the features of the report. Engineering reports written as part of your undergraduate engineering education need not include the executive summary and descriptive abstract, unless required by the instructor.

The table of contents (TOC) is familiar to you, as this text as well as books and reports you have read all have TOCs. But perhaps you never considered the organization of them. The page numbers that start the various sections are noted, as are the topics, or headings, included in the report. Be certain to check that the headings match those in the report, as your revisions may include heading title alterations. The TOC should fit neatly on the page without just a few lines flowing over to the next page, and it should look visually interesting, not congestedly tight or spaciously loose.

Figures and tables are traditionally located at the end of the report, not interspersed throughout. A general guideline is to locate tables of data and technical drawings that distract from reading comprehension in the appendixes at the end. When there are only a few figures or tables, they may be included within the report text. This is comparatively easy to accomplish with word processing software to cut and paste graphs, tables, and drawings. If this is the situation, the figure should be located as near to its mention in text as possible. Also refer to the figure or table in the text, and label as such; you will notice that in the design of this text, the text figures have numbers and labels and are located near to where they are mentioned in the writing.

The report introduction clearly describes the report's purpose and contents. The first paragraph will indicate the specific topic of the report, what it accomplishes and does not accomplish, and the situation that brought about the need for the report. It also includes background information to set the stage for the report, to interest the readers and enable them to understand the context. For long reports, the introduction can explain the organization of the report as well.

The body of the report includes several elements: procedure, results, discussion, conclusions, and recommendations. The procedure section is crucial and because of its technical nature has the potential to be less readable. The optimum solution to the problem—why it is the best solution—is justified as well as the testing and analysis techniques used. It is here that the assumptions used in the analysis are stated and justified, the experimental procedure explained, or the structure and methodology of computer programs reviewed. The details of all these topics, the actual computer program, and analysis sample calculations are placed in an appendix. This section demonstrates your technical prowess, allowing the reader to assess the validity of your assumptions.

The results and discussion of them are usually combined, except in large reports. There will be results from your analysis. More important than the results is your interpretation—are they reasonable in light of your analysis? Error analysis is included here, as there are errors associated with any measurement, with many computer programs, and with statistical sampling of data. No assessment, no value judgment, has been made regarding the results; the next section deals with this critical issue.

In the conclusion section, you apply your engineering know-how to explain why the results are as they are, combining your theoretical knowledge with the actual facts to reach meaningful judgments. In this section you also summarize the key points and key facts that have been discussed, leaving the reader with the perspective you desire. It is also possible to generalize in this section, leading to implications and future developments. Recommendations may be found in a separate section, or they may be included in the conclusion section. In the case of product development reports, questions to be answered include whether a company should go ahead with product and how well it will function over its lifetime. Laboratory reports seldom have recommendations associated with them, as they are concerned with testing and evaluation.

The appendixes follow the end of the report, typically starting with illustrations and tables of data and then the bibliography. It is here that the computer program, the tables of experimental data, or sample calculations are located. There is no hard-and-fast rule about what should be in the appendix, except that the report should read smoothly with anything that might detract from its readability located in the appendix. The bibliography includes text references, handbooks, and journal articles that were used in preparing the report.

Patents, Copyrights, and Trademarks

One of the reasons people consider engineering is that they like inventing things, creating new solutions to solve problems. This is a very rewarding and exciting part of being an inventive engineer; in addition, some ideas are patentable. Patents may be awarded for ideas that may be considered new and that would not be developed by an expert in the routine practice of the profession. This prevents patents for ideas that would impede the profession, such as a patent on fundamental laws. An idea must be new and not routine. A design engineer will routinely use certain techniques—circuits, mechanisms—to solve problems, and these cannot be patented. The importance of patents was noted in the

Constitution where Congress is given the power to enact laws related to same (Article I, section 8).

A patent may be obtained by whoever "invents or discovers any new or useful process, machine, manufacture, or composition of matter or useful improvement thereof." *Process* refers to methods of accomplishing something such as metalworking, refining, constructions, and manufacturing methods, while *machine* refers to a complex mechanism as well as a device having few, or no, moving parts. The *composition of matter* refers to mixtures of ingredients and chemical compounds. In the case of mixtures, the result must be more than the simple effect of multiple compounds, but some new effect resulting from their combination.

The purpose of the patent system is to encourage inventors to disclose their designs to the public in the hope that new technologies can advance with this knowledge. If an inventor does so, the government provides a limited monopoly for 17 years from the time the patent was granted by the U.S. Patent Office. Of course, it is not terribly easy to obtain a patent, as the inventor must demonstrate that it is unique and not described or used by others in the United States or a foreign country. There are patent attorneys, often with an engineering background, and patent agents, trained engineers, who assist inventors to obtain patents. Both facilitate the process, how to fill out applicable forms and follow the necessary procedures, but only the attorneys can argue infringement cases. The patent agent can also provide advice on how to make the patent application stronger and more encompassing.

The patent document discloses the invention in the *specification* and *drawings* and defines the scope of the monopoly in the *claims* section. The monopoly is actually the right to exclude others from using the invention; it does not enable you to produce the device. Say that you invented a new transmission system for a bicycle. You could prevent bicycle manufacturers from using it, although it would be unlikely that you would enter the bicycle manufacturing arena. Normally, you would license a company to manufacture and use the transmission, and in return you would be compensated. People may contest your patent, claiming that it was covered under their previous patents or patents of others that have expired, and hence is not unique. If a company uses the patent without permission, the inventor needs to hire an attorney and litigate, a necessary process to protect the invention.

Initially it may be a little discouraging to find out how many devices and processes have been invented. Many talented people have been creating devices for many years. However, with experience, you may begin to find unique solutions to new problems based on new technologies or materials. Examining areas where new patents have been awarded indicates where the creative energies in engineering are being placed; Thomas Edison did this to seek new opportunities for creating a better device than the one receiving the patent.

This text is not patented, nor is a song; they receive a copyright which protects the author against copying of the material. The copyright law protects the form of expression rather than the subject matter of the writing. For instance, a description of a pencil sharpener can be copyrighted as writing, and others cannot use this description without permission; but they can create their own description of the pencil sharpener.

A trademark relates to any word, name, symbol, or device which is used in trade with goods to indicate the source or origin of the goods and to distinguish the goods from others. The trademark rights may be used to prevent others from using a confusingly similar mark, but does not prevent others from making the same goods or selling them under a nonconfusing mark.

For further information about patents, consult the Web page maintained by the U.S. Patent and Trademark Office at www.uspto.gov. There are forms that can be downloaded, as well as the application process for patents, copyrights, and trademarks.

Doing It

When it is time to write a report, employ the aforementioned style and format. This is the easy part. The more difficult part is doing it. Do not wait until the last moment, of course, and use a word processor, as several drafts will be necessary. Before you begin the writing process, think about the general theme and write it down. Can you explain to someone in a few sentences what you are attempting to convey? This focuses your thoughts. Writing is a creative process, not dissimilar to the design process, in which we need to engage the creative right side of the brain and the analytical left side at different times, recognizing that the analytical side will squelch the creative side if we are not careful. In writing, this can lead to dull, unimaginative thoughts. Jot down ideas you have on a piece of paper; as with brainstorming in design, do not edit the thoughts—let big ideas and small ideas flow. Use only as much notation as required to remember the thoughts later when you organize them. Organize your ideas into a coherent pattern, thinking about how they support the main topic. As you are organizing the ideas, new ones will appear; jot these down and organize also. As you now switch to the creative mode and begin to write, there is a focus as the key points are organized. Do not be overly concerned with spelling and syntax and paragraph structure, this is the creative rough draft. You will have an opportunity, using your analytical understanding of grammar, to improve upon the initial draft. Have a dictionary handy; all writers need one. This is not only for spell checking, which word processors perform, but also to look up words you think you know, to make sure they have the desired meaning. The thesaurus program on computers can assist in this regard also, so that you do not use the same words over and over. Edit the rough draft; professional writers typically write and edit several drafts, and then print out the final report.

References

1. Burghardt, M. David. *Introduction to the Engineering Profession,* 2nd ed. New York: HarperCollins, 1995.
2. Michaelson, Herbert B. *How to Write and Publish Engineering Papers and Reports,* 2nd ed. Philadelphia: ISI Press, 1986.
3. Morris, George E. *Engineering—A Decision-Making Process.* Boston: Houghton-Mifflin, 1977.
4. Strunk, W., Jr.; and E. B. White. *The Elements of Style,* 3rd ed. New York: Macmillan, 1979.
5. Turabian, K. L. *A Manual for Writers of Term Papers, Theses and Dissertations.* Chicago: University of Chicago Press, 1967.

Problems

3.1. Prepare a 5-min talk on an area in engineering of interest to you. Develop the notes and charts (transparencies, flip charts, slides) for the presentation, and give a mock presentation to some classmates. Critique one another.

3.2. Select a technical magazine or journal article, and write an abstract for it.

3.3. Select a consumer product that you would like to test, and develop the criteria and methodology for testing and determining the best one. Then perform the evaluation, noting any changes in your methodology as you were faced with making it scientifically valid. Use either the design portfolio or the design report to document your findings.

3.4. Analyze the design portfolio (particularly after using same for a design project). Are there changes that you would recommend to have it fit with the design process you used? Justify the changes.

3.5. The assessment of design projects and the accompanying documentation are important; analyze the design assessment in the text. Are there any changes you would recommend? Justify the changes.

3.6. Discuss in a short essay the process for obtaining a patent. What difficulties may be encountered in obtaining one?

3.7. What are the differences between patents, copyrights, and trademarks?

3.8. Discuss the differences between invention and innovation.

3.9. Analyze and critique the way in which you write a report or long essay. What are some ways that you can improve what you write?

3.10. Write a letter to your instructor, outlining your expectations for the course.

3.11. Change the passive voice to the active voice in the following sentences.

(a) The scandal was revealed by the company's treasurer.

(b) The manager was suspended by the president.

(c) The company was told by the Department of Environmental Protection to install a filter on two chimneys.

(d) Assemblers should always change stations if they are told to do so by efficiency experts.

(e) A new plan for employees who have been fired is being worked on by the personnel office.

(f) Five different candy bars were recalled by the company.

(g) The lacquer should be spread evenly over the boards.

(h) Almost all keyboard problems can be resolved by special software.

(i) The party always is given by the dean.

3.12. Improve the following sentences, which have unclear modifying words or phrases.

(a) We bought the soda in a small store that cost a dollar.

(b) When covered with flowers, I find the garden beautiful.

(c) After fixing her motorcycle, the contest continued.

(d) We began the new system completely unaware of mechanical problems.

(e) David visited Steve when he was in the hospital.

(f) We ran 10 laps after the workday which exhausted us.

(g) While hidden in the long grass, the lawn mower almost ran over the frog.

(h) Betsy notified Susan she had been told to move to a subordinate position in the company.

3.13. Remove unnecessary words from the following sentences.

(a) In the case of the industrial engineering major who wishes to become an executive, he or she will be expected to have experience at all levels of the company.

(b) There have been many times that I have driven to campus and have spent a half an hour trying to locate a good parking space for my car.

(c) In my opinion, there are several considerations or factors that we on the committee must or should decide so that a decision on the subject will be made.

(d) When he was writing or typing a feasibility study of a projected shopping mall, an area offering a wide variety of stores, he came across new zoning laws that cast doubt on the availability of the already selected building location.

(e) In the time of year we call spring, she drove her car to the automobile dealer who sold cars to replace the catalytic converter located on the car.

3.14. The following sentences need improvement in parallel structure. For instance, *he should be promoted or fired* is improved as *he should be either promoted or fired.*

(a) I was not only interviewed by the chief engineer but by the vice president as well.

(b) This is neither to our advantage or disadvantage.

(c) It is better to give than receive.

(d) We will beat them in the air, land, and sea.

(e) Early to bed and rise makes a person healthy, wealthy and wise.

(f) Families often do not realize how much college will cost and they have no idea of ways to pay for it.

(g) Because we wanted the sales and need them, we all worked overtime.

(h) Whether the tickets are expensive determines if we go.

3.15. Improve the following sentences, which lack unity, often combining several ideas. You will have to determine the logical relationship that the author intended.

(a) David was late to school and brought his lunch with him.

(b) The fool could not see the forest for the trees, but there was a silver lining somewhere.

(c) Sue called Joan to pick her up for school and the car would not start.

(d) After the retirement plan is set in motion works well for the employees.

(e) The students were determined to learn to do computer programming was a year-long task.

(f) When the teacher asks the students to pay attention it is a reasonable request.

(g) The first step is the worker starts the machine.

(h) Although Juan was enrolled in electrical engineering felt he had not received enough math.

3.16. Commas are often misused; insert commas where required and eliminate unnecessary commas.

(a) Middlesex North Dakota is a community that time forgot.

(b) She was born in January, 1951 and shortly afterward her family moved from Hoboken New Jersey to Windsor Locks Connecticut.

(c) His term of office as mayor will expire on 1, January, 1996.

(d) During one year, I lived in Wishing Well, Montana and, for a few weeks, in Sacramento, California.

(e) Alice Lee PE has been practicing engineering for 10 years.

(f) I believe however that the delay is required.

(g) After touring the assembly line robots were all he ever talked about.

(h) John fidgeted in his chair during the trial although he appeared calm but one juror noticed.

4

Engineering Analysis and Design

Analysis and design are linked closely in engineering. When we examined the time line of history, you may have noticed that the rapid increase in inventions and our understanding of the physical world are, in large measure, a result of our being able to analyze them. This is particularly so in more complex systems and devices where intuition, formed from the knowledge base of analysis, has led to these discoveries and inventions. Also, we have noted that innovation, a necessary element in design, requires an understanding of what is happening and why it occurs, so that improvements may be made to the initial design. In this chapter we will examine the fundamentals of analysis in a variety of engineering disciplines and then show a critiqued student design projects that evolved from this knowledge base.

Engineers like to solve problems; sometimes the problems require the design of a new device, at other times the analysis of an existing device or situation. Much of engineering study is devoted to the analysis of problems, be they electric circuits, physical structures, or chemical reactions. There are techniques that you can use that will assist in all these areas, which are ways of thinking and recording your analysis. At times this will take the form of doing homework problems in a particular fashion which will later help you in engineering practice, where a similar recording of your analysis is necessary for others to review. In school you will use your homework problems when reviewing for tests and as references in later courses. You must include sufficient detail to make them both useful to you when the material is not fresh in your mind and understandable to the instructor who is reviewing them.

Figure 4.1 illustrates one type of solution format that is traditionally used. There are various reasons for the following steps, which essentially come down to having you internalize the

Engineering Analysis

problem and apply all your abilities to solve it. Rewriting the problem statement is the first level of this, as you begin to think about what the problem is requiring. Similarly the *given* and *find* sections require that you state what the given information is and ask yourself what is required, both of which force you to think about the problem, not simply restate it. The *sketch and data* section

A spring scale is used to measure force and to determine the mass of a sample of moon rocks on the moon's surface. The springs were calibrated for the earth's gravitational acceleration of 9.8 meters per second squared (m/s^2). The scale reads 4.5 kg, and the moon's gravitational attraction is 1.8 m/s^2. Determine the sample mass. What would the reading be on a balance scale?

Given:
Reading of spring scale weighing mass on the moon.

Find:
Sample mass.

Sketch and Data:

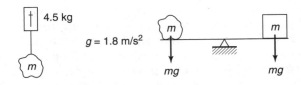

Assumptions:
None.

Analysis:
Determine the force equivalent to a scale reading of 4.5 kilograms (kg) at a gravitational acceleration of 9.8 m/s^2.

$F = mg = (4.5 \text{ kg})(9.8 \text{ m/s}^2) = 44.1$ newtons (N)

Find the mass that will exert a force of 44.1 N under an acceleration of 1.8 m/s^2.

$$F = mg$$

$$44.1 \text{ N} = (m)(1.8 \text{ m/s}^2)$$

$$m = 24.5 \text{ kg}$$

With a balance scale, reference mass and measured mass are both subjected to the same gravitational acceleration; therefore, the reading will be

$$m = 24.5 \text{ kg}$$

Figure 4.1
Sample homework
solution format.

enlists your visual intelligence in the solution. The purpose is to conceptualize what is happening and translate the symbolic word problem into sketches and diagrams. Inherent to the solution of any problem are *assumptions*, such as "the flow is steady," or "the device is in static equilibrium." The assumptions can be listed initially before you begin the analysis section and will often be added to as the analysis evolves. Last, the *analysis* section is where you endeavor to work with the governing equations and relationships to solve the problem.

As you proceed with the analysis section, show the units on your answers, such as amperes or newtons, and clearly indicate your calculations so the instructor knows what you are doing. In later practice when your supervisor reviews your analysis, similar clarity will be needed. The review process is part of virtually all organizations' quality control and reduces the chances of a calculation error or conceptual error. When you have arrived at an answer, clearly indicate it; underline or circle it, and check its reasonableness. Is it the same order of magnitude as other terms? Do not expect that your first attempt will be presentable; in fact, this restricts your creativity, a necessary element in the analysis of problems. Furthermore, the clear presentation you provide will enable the instructor to lend assistance if you cannot complete the problem, pinpointing where the difficulty lies.

Practice in problem solving will develop your engineering intuition. It is discouraging to have numerical results that seem wrong; you end up doubting your solution procedure when perhaps only a calculation error occurred. Minimizing the number of problems on a sheet assists when you study the material later; problems running to another sheet and requiring constant turning of pages can sometimes cause loss of thought continuity.

The analysis portion of the problem solution is where the greatest difficulty for students lies. It is fine to say "Use the appropriate equations to model the solution and solve them," but it is often quite another thing to do this. The key concept in problem solution is the *connection* between the quantities in the problem statement. Problems always have known quantities and something that is to be found, with the analysis connecting them. For instance, given the voltage drop across a known resistor, find the current flow through the resistor. This requires knowledge of Ohm's law; this law provides the connection between voltage, resistance, and current. In class you will study the law and its implications, and you need to remember the variables associated with it. This requires an entirely different mind-set than memorizing equations to substitute into for certain types of problems.

Many, perhaps most, students have difficulty with problems not because of the complexity of the mathematics, but because the students do not understand what they have read. It is a lack of reading comprehension. Problem-solving ability has two parts: the information-processing part—how to solve simultaneous equations; and the information-gathering part—comprehending what the problem is saying. One reason to recommend sketching in problem solution is to ensure that you know what is happening by sketching same. Reading comprehension is essential to good problem-solving ability. Make sure you know what the words mean and imply. Ask yourself, What is happening physically in this situation? Can I explain it? This is often difficult to do, particularly in beginning engineering courses, as the words are symbolic representations of an entity you may not ever have seen, such as a resistor or gas turbine. Seek assistance from your instructor in creating an understanding of devices you do not know so your visualization of the problem can be complete. For instance, density has the unit of kilograms per cubic meter (abbreviated kg/m^3); what does this mean? This is mass (kg) divided by volume (m^3). Solutions to problems involving density will very often require an understanding of mass and volume; mass is invariant and conserved, whereas volume is not conserved and represents the space occupied by the mass.

Introduction to Electrical Engineering

In today's world, we depend heavily on electrical systems for support of our lifestyles and livelihoods. Think for a moment

Electrical engineering students setting up an experiment. *(Courtesy of Hofstra University)*

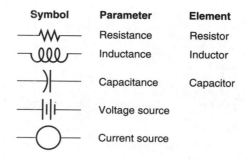

Symbol	Parameter	Element
—\/\/\—	Resistance	Resistor
—ℓℓℓ—	Inductance	Inductor
—)|—	Capacitance	Capacitor
—|‖|—	Voltage source	
—○—	Current source	

Figure 4.2

about all the features of our daily lives that involve electrical or electronic components: waking in the morning to an alarm clock, listening to the radio or watching television, starting an automobile, turning on a light, using a calculator. In every case an electrical system is involved, created and developed by electrical engineers.

All electric circuits can be mathematically modeled by using one or more circuit elements: resistance, inductance, capacitance, voltage source, and current source. Figure 4.2 shows the symbols used to designate these elements. The flow of electric charge requires work to move it from one point to another and is called *current,* a fundamental unit measured in amperes (A). One ampere (1 A) is equal to the flow of one coulomb per second (1 C/s). One volt (1 V) is the change in electrical potential between two points when one joule (1 J) of work is done in moving one coulomb (1 C) of charge from one point to another.

Resistance

The definition of resistance is derived from Ohm's law, named after Georg Simon Ohm, a German physicist who discovered that the voltage change V across a resistor, measured in volts, is equal to the product of the electric current i flowing through the resistor, in amperes, and the resistance R measured in ohms (Ω):

$$V = iR \qquad (4.1)$$

Equation 4.1 may be solved for resistance $R = V/i$. Since this is a law, it cannot be proved, only corroborated by experimental evidence. As with most empirical laws, there are some minor exceptions, but none of note at this level.

The power P in watts (W) that is dissipated in the resistor is

$$P = Vi \qquad (4.2)$$

This, if combined with Equation 4.1, yields

$$P = i^2R \qquad (4.3)$$

In any resistor, the power loss is referred to as the i^2R loss. All conductors have some electrical resistance, so there are power

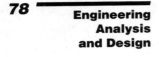

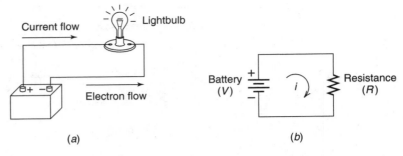

Figure 4.3

(a) (b)

losses associated with current flow through virtually all conductors. In electrical distribution systems, this is very important; if you wish 100 megawatts (MW) of power to be transmitted 100 miles (mi), then you may need to provide 110 MW at the beginning of the transmission line to receive 100 MW at the end.

Steady-State DC Circuits

In steady-state direct-current (dc) circuits, the voltage and current values do not change with time. This is typified by a battery and resistor circuit, shown in Figure 4.3. In the battery a chemical reaction converts chemical energy to electric energy. This process moves electrons, negatively charged, from material at the positive terminal (anode) to the negative terminal (cathode), creating an electric potential. If a load, in this case the lightbulb, acting as a resistor, is connected between the terminals, the electrons will flow from the negative terminal to the positive terminal. The diagram indicates that the conventional current flow, which is hereafter called the *current flow*, is in the opposite direction. This is the direction in which positive-charge carriers would move if the current moved from positive to negative terminals. The movement of positive carriers is important in semiconductor devices, where these carriers are called *holes*. The reason for this inverted direction of current flow is historical. Benjamin Franklin proposed that electric current flow was the flow of positively charged particles. Although it was later proved wrong, the proposal remains and is accepted practice.

Kirchhoff's Laws

Two laws discovered by Gustav Kirchhoff in the mid-1800s are extremely valuable in analyzing electric circuits. The first is frequently called *Kirchhoff's voltage law*:

The sum of the voltage rises around a closed loop in a circuit must equal the sum of the voltage drops.

The second of Kirchhoff's laws is sometimes called *Kirchhoff's current law*:

The sum of all currents into a junction (node) must equal the sum of all currents flowing away from the junction.

The first law assists us in the analysis of resistors connected in series, as shown in Figure 4.4. Three resistors are connected to a dc power source. The circuit has one voltage rise, the battery with voltage V, and three voltage drops, the iR drops across each resistor. Expressing this in an equation yields

$$V = iR_1 + iR_2 + iR_3 \quad \textbf{(4.4)}$$

Figure 4.4

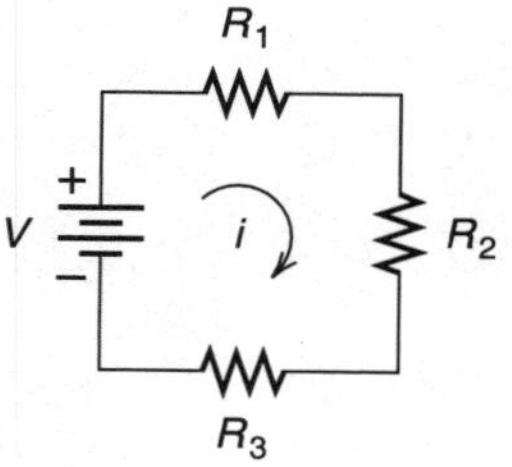

The current flow i is the same for each resistor, hence

$$V = i(R_1 + R_2 + R_3) = iR_{eq} \quad \textbf{(4.5)}$$

where R_{eq} is the single equivalent resistance that could replace all three individual resistances. An equivalent resistance is equal to the sum of the individual resistances.

$$R_{eq} = \sum R_i \quad \textbf{(4.6)}$$

The symbol $\sum$ denotes the summation of the subscripted variable R_i. The equivalent resistance is often used in testing circuits where you are interested in the total resistance load characteristics, not the resistive load characteristics of individual resistors.

In series circuits, the voltage drop across each resistor will vary according to the value of R. For resistors connected in parallel, the voltage drop across each resistor is the same and equal to the battery voltage (Fig. 4.5). From Kirchhoff's voltage law we determine

$$V = i_1R_1 = i_2R_2 = i_3R_3 \quad \textbf{(4.7)}$$

From the current law for junctions x and y we find

$$i_x = i_1 + i_y \quad \textbf{(4.8a)}$$

and

$$i_y = i_2 + i_3 \quad \textbf{(4.8b)}$$

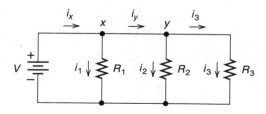

Figure 4.5

Combining Equations 4.8a and 4.8b, we have

$$i_x = i_1 + i_2 + i_3 \qquad (4.9)$$

This makes sense physically in that the current leaving the battery equals the sum of the individual resistor currents. The individual currents may be found by applying Ohm's law to each resistor:

$$i_1 = \frac{V}{R_1} \qquad i_2 = \frac{V}{R_2} \qquad i_3 = \frac{V}{R_2} \qquad (4.10)$$

Substituting Equation 4.10 into Equation 4.9 gives

$$i_x = V \left(\frac{1}{R_1} + \frac{1}{R_2} + \frac{1}{R_3} \right) \qquad (4.11)$$

Notice that the term in parentheses is the inverse of the equivalent resistance R_{eq} for a parallel circuit and that Equation 4.11 is an expression of Ohm's law:

$$\frac{1}{R_{eq}} = \frac{1}{R_1} + \frac{1}{R_2} + \frac{1}{R_3} \qquad (4.12)$$

This may be reduced to a single term:

$$R_{eq} = \frac{R_1 R_2 R_3}{R_1 R_2 + R_2 R_3 + R_1 R_3} \qquad (4.13)$$

We can use the previously developed equations to simplify circuit diagrams.

Example 4.1 Determine the equivalent resistance of the circuit shown in Figure 4.6a.

Assumption:
Steady-state dc circuit.

Analysis:
Start with the resistance(s) farthest from the voltage source. In Figure 4.6a the 15-Ω and 20-Ω resistors are in series and can be combined to form one 35-Ω resistor, as shown in Figure 4.6b. At this point the 10-Ω and 35-Ω resistors are in parallel and can be combined to give an equivalent resistance.

$$\frac{1}{R_{eq}} = \frac{1}{10} + \frac{1}{35}$$

$$R_{eq} = 7.78 \, \Omega$$

The circuit diagram now is shown in Figure 4.6c. The two remaining resistors are in series and can be added, as shown in Figure 4.6d, yielding an equivalent resistance for the circuit of 12.78 Ω.

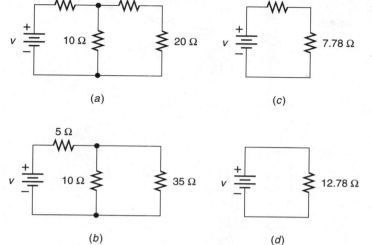

(a)

(c)

(b)

(d)

Figure 4.6a–d

Two applications that we will examine for purely resistive circuits are the variable voltage divider, also known as a *potentiometer*, and the Wheatstone bridge. A potentiometer is a rheostat, a device whose resistance can be varied (Fig. 4.7). One of the leads of the resistive material is attached to ground, with leads of the wiper and the other end connected to the voltage source. Figure 4.8 illustrates a potentiometer in a circuit; the potentiometer's wiper has been connected to a loudspeaker. The variable resistance changes the voltage supply v across the speaker, thereby changing the volume. The lowercase v is used for the variable voltage supply of the music coming from the amplifier. In this situation the voltage is applied across the potentiometer's total resistance, with the loudspeaker connected between the wiper and ground. In Figure 4.8, the speaker receives only 20 percent of the voltage drop, and the volume is low; in Figure 4.9 the wiper—the volume

Applications of Resistive Circuits

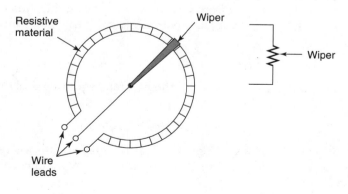

Figure 4.7

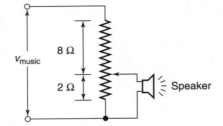

Figure 4.8

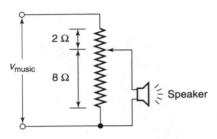

Figure 4.9

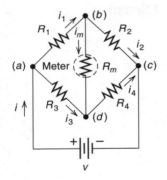

Figure 4.10

knob on the stereo—has been turned to allow 80 percent of the voltage drop to occur across the speaker, and the volume is higher. Variable-resistance potentiometers are used as volume and tone controls on radios and stereophonic equipment and as contrast and brightness controls on televisions and monitors.

The Wheatstone bridge is a special case of variable resistance, included in a parallel and series connection of resistors, that can measure very small values of resistance, such as those found in strain gages. Strain gages are used in testing the strength of materials to find the load that a material is subjected to and the effect of that load on creating a strain in the material. Figure 4.10 illustrates a Wheatstone bridge with current flows in the resistors for all the branches. There are resistances in each of the arms and a meter with a resistance R_m. In practice, the R for one arm is unknown and the bridge is balanced to find the unknown resistance, frequently the strain. Let us consider the bridge is balanced; thus i_m, the current through the meter, is zero. The current in the different branches must be equal, thus $i_1 = i_2$ and $i_3 = i_4$. Furthermore, the potential across the meter must be zero, thus

$$i_1 R_1 = i_3 R_3 \quad \text{and} \quad i_2 R_2 = i_4 R_4$$

Eliminating i_1 and i_4 from the previous equation and dividing the result yield

$$R_1/R_2 = R_3/R_4 \qquad \textbf{(4.14)}$$

In a Wheatstone bridge, R_1 and R_2 may be known resistances, R_3 a variable resistance, R_4 an unknown resistance, and R_m the resistance of a galvanometer, which indicates the current i_m in an unbalanced bridge arm. The variable resistor is adjusted to bring i_m to zero. The final resistance of R_3 is read from a dial attached to it, and the value of the unknown resistance is determined from Equation 4.14.

Another use of a balanced bridge is found in an alarm system. Consider the circuit in Figure 4.10 with resistances balanced per Equation 4.14; in place of the meter, substitute an alarm which activates if current passes through it. Now one of the resistances is a thermistor—a resistor whose resistance changes with temperature. One resistor of the bridge unit is located in a refrigerated space. If the compressor fails and the temperature rises beyond a certain point, allowing a sufficient change in resistance, the current flow will be great enough to trigger the alarm. Note that changes in voltage do not affect the alarm.

The circuits described above are totally resistive and hence are not transient. There is no delay in current or voltage in a resistive circuit; all changes occur instantaneously. However, other circuit elements, capacitance and inductance, have properties that delay changes in current or voltage.

Computers

Computers characterize our current society in much the same way as factories characterized the earlier industrial era. When one thinks about the industrial era, images of factories and smokestacks emerge. The postindustrial era, often called the computer era, is characterized by computers influencing many aspects of our lives, the types of services available to us, and the technologies we use.

Digital computers use binary numbers in performing the various calculation procedures. Information is stored in registers—devices that have two states that the computer reads as *on* or *off*; hence, the binary system fits with computer's functioning (see Appendix A for information on number systems). Each digit is referred to as a bit (*bi*nary dig*it*), and a group of bits forms a word. A word may contain a certain number of bits—8, 16, 32—which are also known as *bytes*. In Figure 4.11 an 8-bit word is visualized as eight compartments containing 1 bit. The most significant bit is the one farthest to the left, whereas the least significant bit is farthest to the right. The computer program that

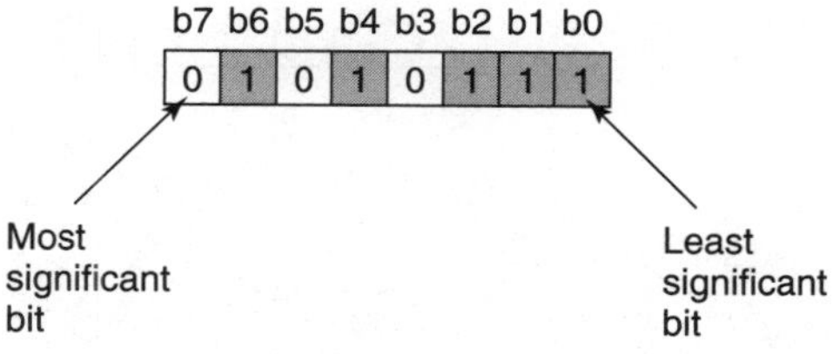

Most significant bit

Least significant bit

Figure 4.11
An 8-bit word, or byte.

uses the words must determine the meaning of the bit location. The same word could mean a letter in the alphabet, a number with its sign, or instructions to the computer to perform a task. Because each location can have only an on or off value, it is adaptable to being magnetized or not. A magnetic recording head determines whether a bit location is magnetized (1 or 0), and this information is transferred for use in the computer.

Logic Diagrams

Logic diagrams illustrate the path of information within a computer. Computer engineers design logic diagrams using three fundamental elements: the AND gate, the OR gate, and the inverter. Other circuit elements may be derived from these. The gates may be constructed physically from a wide variety of electric and electronic switches, diodes, transistors, or fluidic devices. Whereas the same logic diagram can be used regardless of the physical device, the computer engineer must judge the speed of signal transmission, cost, and availability of the device in specifying the actual circuit construction. In computers the gates are transistors, contained within an integrated-circuit (IC) chip.

AND Gate

The AND gate is a device whose output is a logic 1 only if both inputs are logic 1. If one input is logic 0, then the output will be logic 0. Logic 1 and logic 0 states are, by convention, known as *closed* and *open*, *high* and *low*, or *true* and *false*.

Consider the simple electric circuit shown in Figure 4.12*a*. A battery is connected to a light, which may be on only if both switches *A* and *B* are closed. If either is closed without the other, no current can flow, and the light will be off. Figure 4.12*b* symbolizes this AND gate, and Figure 4.12*c* provides the truth table and logic equation. The table shows that the only way for the light to be on is the logic 1 state. The Boolean logic equation

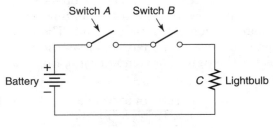

Figure 4.12a
The electric circuit analogous
to the logic AND gate.

Figure 4.12b
The AND gate logic
symbol.

Truth table

Input		Output
A	B	C
0	0	0
0	1	0
1	0	0
1	1	1

Logic equation:
$$C = A \cdot B$$

Figure 4.12c
The AND gate truth table and logic equation.

$C = A \cdot B$ or $C = AB$ is read "C equals A and B," meaning that for C to be logic 1, both A and B must be logic 1.

OR Gate

The electric circuit analogy for the OR gate is shown in Figure 4.13a with the accompanying OR gate symbol and truth table in Figure 4.13b and 4.13c. In this situation the light will be on (logic 1 state) if either switch or both switches are closed. The truth table reflects these conditions. The logic equation is read "C

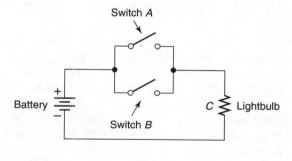

Figure 4.13a
The analogous circuit for the OR gate.

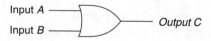

Figure 4.13b
The OR gate logic symbol.

Truth table

Input		Output
A	B	C
0	0	0
0	1	1
1	0	1
1	1	1

Logic equation:
$$C = A + B$$

Figure 4.13c
The OR gate truth table and logic equation.

Figure 4.14a
The inverter logic symbol.

Input A ———▷○——— Output C

Truth table

Input	Output
0	1
1	0

Logic equation:
$$C = \bar{A}$$

Figure 4.14b
Inverter truth table and logic equation.

equals A or B." Notice that the symbol + is the Boolean symbol for OR, not addition.

Inverter

The last logic gate we will consider is the inverter, or NOT gate. It simply changes the input state from logic 1 to logic 0, or vice versa. It is shown symbolically in Figure 4.14a and with the accompanying truth table and logic equation in Figure 4.14b.

Logic Circuits

It is possible to join two or more gates to provide a variety of logic functions. It is not necessary to be limited to two input signals; any number of inputs may be considered. We will consider just three, as illustrated schematically in Figure 4.15a with the truth table and logic equation in Figure 4.15b. For a logic 1 state as the output, we need logic 1 states for the inputs to gate 2, and this is achieved only if logic 1 states are inputs to gate 1.

Consider the combination of an AND gate and an OR gate, creating an AND/OR gate. Figure 4.16a illustrates the logic circuit with the truth table and logic equation in Figure 4.16b. For a logic 1 condition at the output of gate 2, we need a logic 1

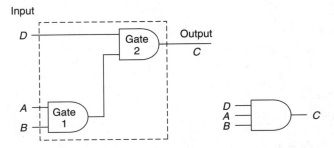

Figure 4.15a
A three-input AND gate.

Truth table

Input			Output
D	**A**	**B**	**C**
0	0	0	0
0	0	1	0
0	1	0	0
0	1	1	0
1	0	0	0
1	0	1	0
1	1	0	0
1	1	1	1

Logic equation:

$C = D \cdot (A \cdot B)$

Figure 4.15b
Three-input AND gate truth table and logic equation.

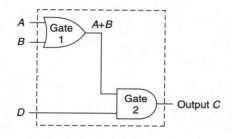

Figure 4.16a
A three-input AND/OR gate.

Truth table

Input			Output
D	**A**	**B**	**C**
0	0	0	0
0	0	1	0
0	1	0	0
0	1	1	1
1	0	0	0
1	0	1	1
1	1	0	0
1	1	1	1

Logic equation:

$C = (A + B) \cdot D$

Figure 4.16b
Three-input AND/OR gate truth table and logic equation.

state for the D input and for the output of gate 1. To achieve a
logic 1 state for the output of gate 1, we need either A or B inputs
to be logic 1; the output, state C, is logic 1.

A challenge was made to a group of first-year engineering students to create a game that used elementary electric circuits. The
focus is on the design process, not the complexity of the circuit.

**Electrical
Engineering
Design Report**

TEACHER IN A BOX

Joseph DiBiasi
Introduction to Engineering
Section 03

A GAME USING ELECTRIC CIRCUITS—TEACHER IN A BOX

Introduction

"Teacher in a Box" is the solution to the design challenge to create a game or device that uses elementary electrical circuitry. This electronic game challenges people of any age to help in the learning process. Questions are placed on a punch card, and the card is inserted into the machine. The user then presses a button corresponding to the three choices given on the card, *A, B,* or *C*—and if the correct letter is pressed, a green light comes on; if an incorrect letter is pressed, then the light does not go on.

Research and Investigation

In addition to creating a game using series and parallel circuitry, the following specifications were added: The game should be engaging to children; the game should be portable; it should use batteries and operate a long time without changing batteries; and it should be inexpensive to build.

Before brainstorming for a variety of possible solutions, an investigation of two toy stores was performed, looking at games being sold to children. A holiday catalog from a department store was scanned for possible toy ideas, and two children who live next door were questioned as to what they like in a toy. The investigation yielded information that many toys exist, but they are complicated to fabricate and often seem to use microprocessor chips. Children indicated they liked toys that reacted to what they did, a light going on or sounds emitting or both. In brainstorming for ideas, several came to mind concerning motion and movement; others focused on lights and buzzers in response to questions being asked. The first group of ideas was not used because they were too complicated. By thinking more about the second group, the idea emerged of having questions answered and when the correct answer is chosen, a light shines.

Further modification of the second idea yielded the optimum configuration, shown in Figure 4.17*a*. In this design, there is a light that goes on when there is a correct answer, and no light appears for an incorrect answer. It is possible to have a buzzer sound as well by adding it in series to the light. A decision was made to use a light-emitting diode (LED) as the light source, rather than a lightbulb, to signal the correct answer. A lightbulb requires only 1.5 volts (V), whereas an LED requires 3.0 V, but the LED will last longer and is colored, creating a more festive appearance. Also, the wiring is simpler, as no bulb holder is required, although a resistor is added to the circuit, as will be explained now.

Analysis

The game should operate a long time without needing new batteries, so an analysis of the power consumption of the circuit was performed. The circuit is really quite simple—a series circuit with the battery voltage supply, two switches, the LED, and the resistor, illustrated in Figure 4.17*b*. An LED requires a certain threshold voltage to start emitting light; until that threshold is reached, the diode effectively blocks the flow of current in the circuit. Further investigation found that the diode resistance varied with the current flow through it. A test was conducted on the diode in which different resistances were placed in series with the diode and connected to two AA batteries. The results are shown in Figure 4.17*c*. The current flow of 4.45 milliamperes (mA) occurred with a 270-Ω resistor in the circuit, the resistor used in the final design. The batteries have a duration of 2450 milliampere-hours (mA · h). At a current flow of 4.45 mA the battery will last

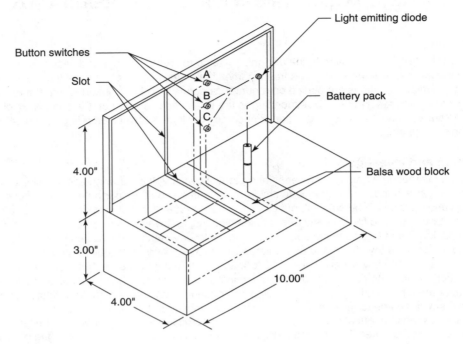

Figure 4.17a
Teacher in a Box.

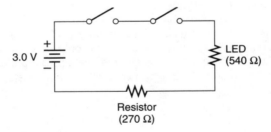

Figure 4.17b

$$2450 \text{ mA} \cdot \text{h} = (4.45 \text{ mA})(t\,\text{h})$$

$$t = 550 \text{ h}$$

It was estimated that each time the correct button was pressed, the light would stay on for one second or 3600 correct answers per hour. Thus the number of times the button can be pressed before battery failure is (550 h)(3600 turns/h) = 1,980,000, or nearly 2 million times. Even if the button were pressed for 2 seconds on average, the game could be played a million times before

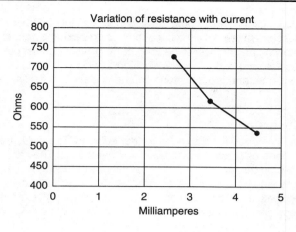

Variation of resistance with current

Figure 4.17c

the batteries needed changing. The requirement that the game be long-lasting is certainly met. The LED lifetime may be dramatically shortened to several hours if a current-limiting resistor is not used. The button switches were tested for their resistance when closed, about 0.1 Ω, or negligible when compared to the LED resistance and that of the resistor.

Construction

There were several additional considerations that arose during the construction phase of the project, modestly impacting the design. Initially a thin wooden box with an open bottom constructed from balsa wood was investigated, but arranging the circuit and the slot upside down was difficult. Finally, a shoe box was selected as the containment device, and it had an attached lid so the wiring could not be pulled apart. Two AA cells were wired in series and connected to the correct polarity of the LED.

Another aspect of the design involved figuring out how to have the card contact the switches in the parallel/series circuit. The switches are those found in video games. They are very sensitive to pressure, so only a slight force is needed to cause them to close, yet they are sturdy, so they will not close if the box is shaken. The card was notched so that the correct answer, for instance, *B*, causes that switch to close and the other switches remain open. As a result of pushing button switch *B*, both switches in the circuit close and the light shines. The button switches are delicate, and it took some soldering dexterity to connect the wires to them. The wiring was reinforced with tape.

The index cards are flexible, so a wooden guide is used to direct the cards to the video game switches. The guide was made from balsa wood and attached to sides of the box, lending additional lateral stability to the box. The index cards had slots cut into the bottom edge, the correct one to engage the switch, gaps to prevent closing incorrect switches, and dummy gaps to mask the correct answer.

Testing

The game was tested with adults and children, and it performed correctly all the time. No incorrect answers caused the LED to light. There are potential difficulties concerning the sturdiness of the cards. If they are inserted with too much force, it will cause the slots to bend and, with continued use, not to close the switches. In an extreme case, one could push the card so forcefully that all the switches are closed, allowing any answer to be correct. To protect against overzealous insertion, a line was drawn across the card, indicating the position to which it should be inserted. The participants all liked the game, and the questions can be written for different age levels and on different topics.

Discussion and Conclusion

Teacher in a Box can be improved if it is to have further editions. The cards need to be made of a more durable material, perhaps plastic or plastic-coated cardboard. The cards should also have something, perhaps a tab on the top, to prevent insertion upside down, which would close all the switches. There also needs to be a midpoint tab or taper to the card that prevents overzealous insertion, closing all the switches. There are only three answers on the card for two reasons: first, three proved the concept; second, there were only three video switches available.

Bibliography

1. Burghardt, M. David. *Introduction to Engineering Design and Problem Solving.* New York: McGraw-Hill, 1999.

2. Serway, Raymond A. *Physics for Scientists and Engineers with Modern Physics*, 3rd ed. New York: Saunders, 1990.

3. Michaelson, Herbert B. *How to Write and Publish Engineering Papers and Reports*, 2nd ed. Philadelphia: ISI Press, 1986.

Assessment Critique

Teacher in a Box is a creative, well-executed project, and the assessment rubrics identify exactly why this is so. What follows is an annotated assessment sheet for the project. There was class discussion as to what the various levels of accomplishment meant for each of the rubrics, called *benchmarking*.

DESIGN ASSESSMENT RUBRICS

The Design Process

A. Identified problem criteria, constraints, and specifications.   0   1   2   (3)
Student expanded on the problem statement, adding unique constraints and specifications such as long life and children's enjoyment of the game.

B. Gathered background information from a variety of sources.   0   1   2   (3)
Several sources—stores, catalogs, and interviews—were used to establish game attributes.

C. Suggested several alternative solutions.   0   1   (2)   3
Good general description of alternative solutions, but there is not the detail to understand them well. A few more sentences adding these details are required for a 3.

D. Evaluated ideas against design criteria and made improvements.   0   1   (2)   3
There was not a detailed description of how other designs met the design criteria; this element tended to be treated lightly in the report.

E. Justified the chosen solution.   0   1   (2)   3
The final solution is indeed the best one presented, but a comparison on a point-by-point basis was not made with alternatives.

The Design Solution

A. Provided an accurate drawing with basic details and dimensions.   0   1   2   (3)
Well-executed CAD drawing with overall dimensions. (If the project is in conjunction with a CAD course, the level of drawing may be higher and more drawing elements may be included in the assessment.)

B. Constructed the model and used materials appropriately.   0   1   2   (3)
The model was fabricated with attention paid to detail. The wood sides and top did not overlap, and the wiring and internal guide were positioned nicely.

C. The solution worked. It fulfilled the design criteria.   0   1   2   (3)
The testing and analysis of the project showed that it solved the design problem with the additional constraints and specifications that were added.

D. Originality and creativity of the design.   0   1   2   ③

This was a very clever design that uses simple electrical circuitry to create an educational game.

Testing

A. Used knowledge gained from testing to inform design.   0   1   2   ③

There were two ways that testing informed the design: First, the LED testing led to placing a resistor in the circuit to maintain long LED life; second, a line was drawn on the card to prevent overzealous insertion in this prototype.

Work Habits

A. Completed assigned task in a timely fashion.   0   1   2   ③

The assignment was turned in when it was due, both the project and the report.

Communication and Presentation

A. Demonstrated understanding of key ideas orally and/or in writing.   0   1   2   ③

The report is clearly written with the design process elements included.

B. Report neatly written with good grammar.   0   1   2   ③

The appearance of the report is excellent, and the grammar—spelling, syntax, and format—is well done.

Scoring guide: 0 = No response or unacceptable response
1 = Acceptable response
2 = Good response
3 = Excellent response

Score: 36
Total possible points: 39

4.1. Find the equivalent resistance for the circuit in Figure P4.1.

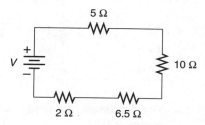

Figure P4.1

4.2. Find the equivalent resistance for the circuit in Figure P4.2.

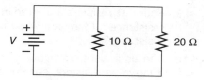

Figure P4.2

4.3. Find the equivalent resistance for the circuit in Figure P4.3.

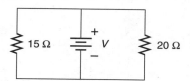

Figure P4.3

4.4. Find the equivalent resistance for the circuit in Figure P4.4.

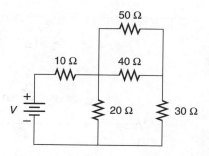

Figure P4.4

4.5. Find the equivalent resistance for the circuit in Figure P4.5 (shown on p. 96) and the current flow through each resistor.

**Engineering
Analysis
and Design**

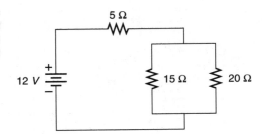

Figure P4.5

4.6. Find the current flow through each resistor in Figure P4.2 if the voltage source is 12 V.

4.7. The filament in a flashlight bulb has a resistance of 50 Ω, and the battery voltage is 6 V. Determine the current flow.

4.8. The maximum current flow from a 1.5-V battery is 45 mA. What is the minimum size resistor that can be connected to it?

4.9. Four 5-Ω resistors are wired in parallel circuit. What is the equivalent resistance? If they are wired in series, what is the equivalent circuit resistance?

4.10. The equivalent resistance of a three-resistor series circuit is 39 Ω. If the three resistors, each of the same value, are now connected in parallel, what is the equivalent circuit resistance?

4.11. Find the equivalent resistance for the circuit shown in Figure P4.11.

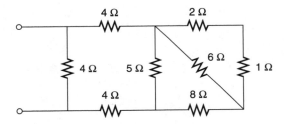

Figure P4.11

4.12. Find the equivalent resistance for the circuit shown in Figure P4.12.

Figure P4.12

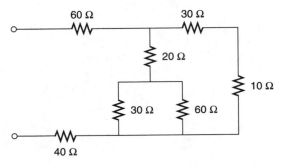

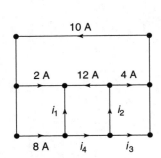

4.13. Use Kirchhoff's current law to determine the unknown currents in Figure P4.13.

Figure P4.13

4.14. Find the power absorbed in the 10-Ω resistor in Figure P4.14 (p. 97).

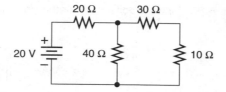

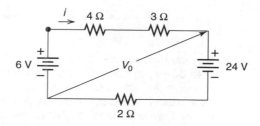

Figure P4.14

4.15. Find the current and voltage for the network in Figure P4.15.

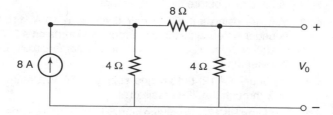

Figure P4.15

4.16. Determine the voltage V_o for the circuit in Figure P4.16.

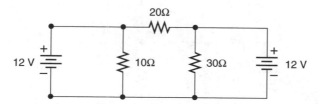

Figure P4.16

4.17. Determine the current i_1 for the circuit in Figure P4.17.

Figure P4.17

4.18. Determine the value of the current source in Figure P4.18.

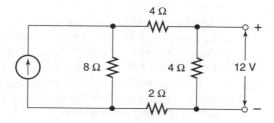

Figure P4.18

4.19. Find the voltage V_o for the circuit network in Figure P4.19.

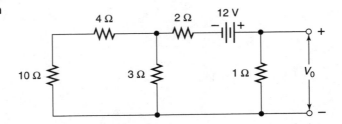

Figure P4.19

4.20. Your car radio is broken, so you are using a 9-V transistor radio that uses 30 mA of current. Being an engineering student, you wish to conserve the radio's battery and want to run the radio from the car's 12-V battery. A resistor must be placed in series with the radio to reduce the car's voltage to that of the radio; what is its value? What power does the transistor radio dissipate?

4.21. Many home lighting circuits have 15-A circuit breakers with a power supply of 110 V. How many 100-W lightbulbs may be placed in parallel in the circuit before the breaker trips?

4.22. You are using a 1250-W hair dryer on a 15-A, 110-V circuit. Your younger sister comes into the room and turns on a 350-W stereo and a 100-W light, also in parallel on the same circuit. Does the circuit breaker trip?

4.23. In Figure P4.12, the voltage supply is 50 V. Determine the current flow through the 20-Ω resistor.

4.24. In Figure P4.12, the voltage supply is 50 V, and an open occurs across the 10-Ω resistor. Find the current flow through the 20-Ω resistor.

4.25. The power produced by the 50-V source in Figure P4.25 is 300 W. Determine R_1.

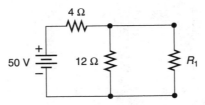

Figure P4.25

4.26. What is meant by *balance* when one is referring to a bridge? How is it accomplished?

4.27. A Wheatstone bridge, as illustrated in Figure 4.10 , is used to determine the value of R_1. The bridge is balanced, and R_3 reads 137.5 Ω. Resistors R_2 and R_4 are interchanged, the bridge is balanced, and R_3 now reads 167.9 Ω. Determine R_1.

4.28. In a Wheatstone bridge, the resistances R_3 and R_4 are equal. An unknown resistance R_1 is connected to the bridge, and the bridge is balanced, yielding a value for R_2 of 400 Ω. What is resistance R_1?

4.29. Are the bridges in Figure P4.29a and b balanced?

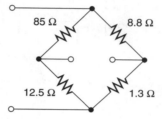

Figure P4.29a

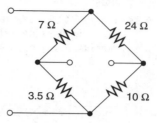

Figure P4.29b

4.30. Prepare a truth table for the circuits in Figure P4.30a and b.

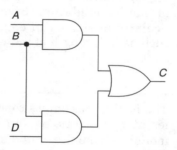

Figure P4.30a

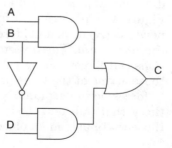

Figure P4.30b

4.31. You have been assigned the task of creating the logic circuit for a bank alarm. The alarm is to sound, logic 1 state, if the master switch is on, if the bank door is open, and if the safe door is open.

4.32. Design the logic circuit for starting an emergency diesel generator. For the generator to start, the generator must be disconnected from the bus, there must be sufficient starting air pressure, and the fuel

**Engineering
Analysis
and Design**

tank must indicate there is oil. When these conditions are met, then the diesel may be started. Presume there are sensors to determine these values.

4.33. Figure P4.33 illustrates a schematic diagram for a half-adder; it will add any two binary digits and produce the resulting summation as outputs. Show that it will correctly add all combinations of binary input.

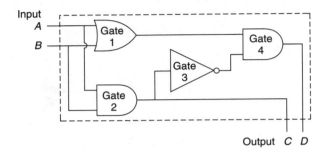

Figure P4.33

Engineering Mechanics

One of the first engineering courses you will take is in engineering mechanics, the first part of which concerns itself with bodies in static equilibrium. Civil and mechanical engineers are particularly interested in engineering mechanics, as later courses, such as in strength of materials and structural analysis, depend on it.

Forces

The field of mechanics concerns itself with forces acting on bodies at rest—static equilibrium—and forces acting on bodies in motion—dynamics. Fundamental to our understanding of mechanics is a knowledge of forces. Forces have magnitude and direction and hence are mathematically represented as a vector. Figure 4.18 illustrates a fixed vector **A**. It has a magnitude of A newtons (N) or pounds force and is acting at an angle of $\theta°$ from the horizontal. Force vectors range from the simplest combinations, all forces acting collinearly as shown in Figure 4.19a, to those acting at the same point or concurrently, as in Figure 4.19b, to forces acting coplanarly, as in Figure 4.19c. Tensile forces are those that pull on a body or object, and compressive forces are those that push on an object.

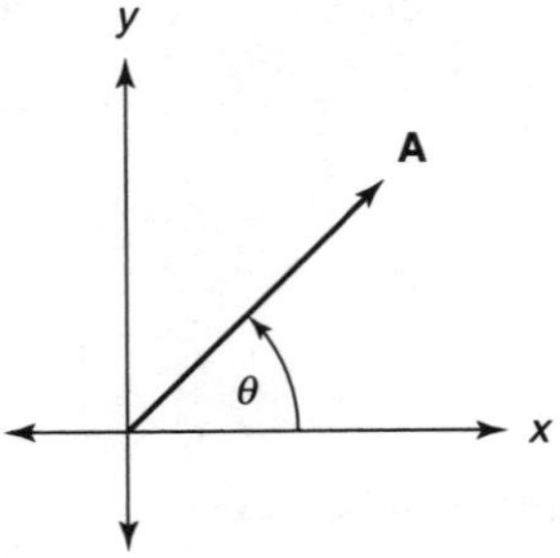

Figure 4.18
A fixed vector.

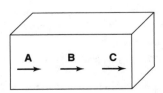

Figure 4.19a
Collinear forces.

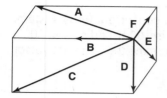

Figure 4.19b
Concurrent forces.

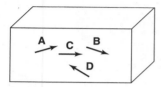

Figure 4.19c
Coplanar forces.

Vector forces rarely occur at neat angles of 90°, and to simplify the analysis, we must be able to resolve forces into x and y vector components.

Example 4.2   A force of 100 N acts at an angle of 30° to the horizontal. Determine its horizontal and vertical components.

Sketch and Data:

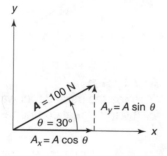

Figure 4.20

Analysis:
From trigonometry we know that

$$A_x = A \cos 30 = (100\,\text{N})(0.866) = 86.6\,\text{N} \rightarrow$$

$$A_y = A \sin 30 = (100\,\text{N})(0.5) = 50.0\,\text{N} \uparrow$$

Very often we determine the resultant of concurrent forces by adding the individual rectangular components of the vector and then determining the resultant vector along the hypotenuse formed by them. The resultant vector can also be determined by carefully adding the vectors tip to tail on graph paper, and the resultant can be determined graphically.

Example 4.3   Three vectors form a concurrent force system. Vector **A** is 50 N and acts at 45°, vector **B** is 100 N and acts at 300°, and vector **C** is 75 N and acts at 150°. Determine the resultant vector (magnitude and direction).

Sketch and Data:

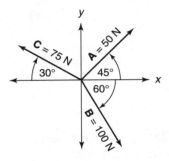

Figure 4.21a

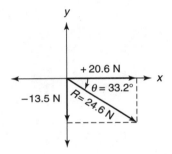

Figure 4.21b

Analysis:

Perhaps the most direct way to solve the problem is to set up a table for the individual components, and then sum them.

Force N	Horizontal component N	Vertical component N
100	(100 N)(cos 60) = +50.0 N	(−100 N)(sin 60) = −86.6 N
50	(50 N)(cos 45) = +35.6 N	(50 N)(sin 45) = +35.6 N
75	(−75 N)(cos 30) = −65.0 N	(75 N)(sin 30) = +37.5 N
	Total +20.6 N	−13.5 N

Figure 4.21*b* illustrates the resultant vector and the resultant found by graphical means. The resultant **R** is found from the Pythagorean theorem, with the angle being determined from the definition of tan θ, sin θ, or cos θ. Thus,

$$\mathbf{R} = (20.6^2 + 13.5^2)^{0.5} = 24.6 \text{ N}$$

and

$$\theta = tan^{-1}13.5/20.6 = 33.2°$$

It is not necessary to resolve two intersecting forces through the summation of components: We may use relationships developed through trigonometric identities and reduce the complexity of the process. Additionally, we will be able to determine the resultant and components. Consider the situation in Figure 4.22a, where two forces **A** and **B** are acting on a tent peg. We can determine the resultant force, using the law of cosines and the law of sines. Figure 4.22b shows the resolution of the forces. Using the general notation of Figure 4.22b, we obtain the magnitudes from the law of cosines:

$$R^2 = A^2 + B^2 - 2AB[\cos \theta(\pi - \theta)]$$

and since $\cos(\pi - \theta) = -\cos \theta$, this may be reduced to

$$R^2 = A^2 + B^2 + 2AB \cos \theta$$

Thus for angles given in Figure 4.22a and by letting $A = 100$ N and $B = 200$ N, the resultant magnitude is

$$R^2 = 100^2 + 200^2 + 2(100)(200)\cos 20° = 87\,588$$

$$R = 296.0 \text{ N}$$

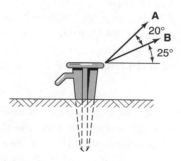

Figure 4.22a
A tent peg with two concurrent forces acting on it.

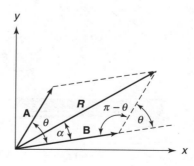

Figure 4.22b
The resultant **R** of two forces **A** and **B**.

If the resultant magnitude is known and the angles at which the components are acting are also defined, then the magnitudes of components A and B may be determined by using the law of sines:

$$\frac{R}{\sin(\pi - \theta)} = \frac{A}{\sin \alpha} = \frac{B}{\sin(\theta - \alpha)}$$

The magnitudes of the components are

$$A = R\,\frac{\sin \alpha}{\sin(\pi - \theta)}$$

$$B = R\,\frac{\sin(\theta - \alpha)}{\sin(\pi - \theta)}$$

Moments

We have thus far considered concurrent forces acting on a body. A moment occurs when coplanar forces act on a body at different points, causing an unrestrained body to rotate. A moment about a point is a force times the perpendicular distance to the point. Consider the wrench in Figure 4.23, with a force $\mathbf{F}$ acting at a distance x. The product of $\mathbf{F}$ and x is the moment about A, or

$$\mathbf{M}_A = \mathbf{F} \cdot x \qquad (4.15)$$

where x is the perpendicular distance from F to A. Moment is a vector with units of newton-meters $(\mathrm{N} \cdot \mathrm{m})$. In this case the wrench would rotate in the counterclockwise direction, by convention a positive moment. Clockwise moments are negative. Notice from Figure 4.24 that the distance must be the perpendicular distance from the line of action of the force to some center of rotation, such as A. In this case, the moment about A is

$$\mathbf{M}_A = \mathbf{F} \cdot x \, \sin \theta$$

Forces that cause moments are often created by an object's weight. Although the weight is not located at one point but is distributed over the entire object, it may be considered to act at one point, its center of gravity. For objects with uniform weight

Line of action →

$\mathbf{F}$

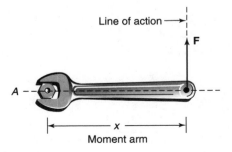

Figure 4.23
A moment $\mathbf{M}_A = \mathbf{F} \cdot x$, created by a force $\mathbf{F}$ acting at a distance x about point A.

A

x

Moment arm

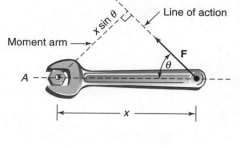

Figure 4.24
A moment $M_A = F \cdot x \sin \theta$, created by a force **F** times the moment arm $x \sin \theta$ about point A.

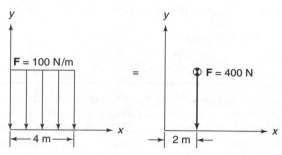

Figure 4.25
An object's uniformly distributed force may be replaced by a single force acting at its center of gravity.

distribution, the center of gravity corresponds to the geometric center of the object. The symbol Θ indicates the center of gravity of an object. Very often a material will have a certain weight, force per linear distance, and can be represented as in Figure 4.25.

Example 4.4   Figure 4.26a illustrates a rigidly attached beam at point C with two forces acting on it at points A and B. Determine the moment about C created by these forces.

Sketch and Data:

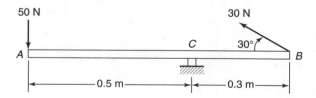

Figure 4.26a

Simplify the problem by finding the horizontal and vertical components of force B. The horizontal component is collinear with C and will not cause a moment to occur. Force A only has a vertical component, and the vertical components of forces A and B will cause the moment. Figure 4.26b illustrates the resolution of forces.

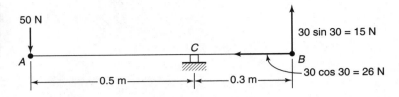

Figure 4.26b

The moment about C is

$$\mathbf{M}_C = (50\,\text{N})(0.5\,\text{m}) + (15\,\text{N})(0.3\,\text{m}) = 29.5\,\text{N} \cdot \text{m}$$

Free-Body Diagrams

At this stage it is important to develop a mathematical model of the body on which the forces and moments are acting. The model is called the *free-body diagram* and is a schematic drawing of that portion of the body under analysis, isolated from other parts, with forces and moments from other parts shown. It is "free" in that the other structural elements are represented only as forces and moments. It should be apparent, but may be forgotten when you are trying to solve a problem, that if the model is wrong, the analysis is useless. Drawing the free-body diagram correctly is an essential first step in mechanics problem analysis.

For example, Figure 4.27*a* illustrates a cargo boom on a ship; the boom may be adjusted through a pivot to position the cargo crate with weight **W**.

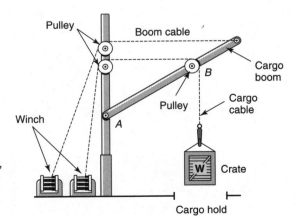

Figure 4.27a
A cargo boom is positioned, raised or lowered, with the boom cable. The cargo is raised or lowered with a separate cargo cable.

The free-body diagram at point B is shown in Figure 4.27*b*.

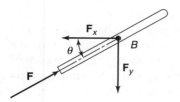

Figure 4.27b
Free-body diagram at
point *B*.

Note that only the forces acting on *B* are shown. The cable can support forces only in tension, and these forces must act along the axis of the cable. The boom can support forces in tension or compression, with the force vector acting along the boom axis. In this situation the downward force $\mathbf{F}_y$ is equal in magnitude to the weight of the crate, and the horizontal force $\mathbf{F}_x$ must be equal in magnitude to $\mathbf{F}_y$. The force $\mathbf{F}$ in the beam must the vector sum of these forces acting in the opposite direction.

When an object is pinned, the pin can support forces in two dimensions, whereas a roller can support only transfer vertical forces—a horizontal force would cause it to roll. Figure 4.28*a* illustrates a load-carrying beam pinned at one end and with a roller support at the other. The free-body diagram with the reactive forces is shown in Figure 4.28*b*.

The reactive forces $\mathbf{R}_1$, $\mathbf{R}_2$, and $\mathbf{R}_3$ are created to counter the imposed load $\mathbf{W}$ on the beam. Free-body diagrams are not an end in themselves, but are used in conjunction with conditions of static equilibrium to find unknown forces acting on a body.

Consider yourself standing still on the floor. You are in static equilibrium. There is a force balance; the downward forces, a result of gravitational acceleration acting on your body mass, are balanced by an equal and opposite force from the ground upward under your feet. This is the weight you feel. It counters the downward force, and a condition of static equilibrium exists. When you fall, static equilibrium is lost until you are firmly on

Static Equilibrium

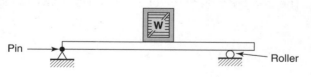

Figure 4.28a
A loaded beam with one hinged end, which supports forces in all directions, and one vertically supported end, which only supports vertical forces.

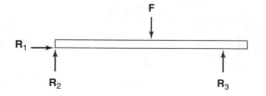

Figure 4.28b
Free-body diagram for the beam.

the ground. Because we are three-dimensional beings, the forces balance in each dimension; we could say that the sum of the forces, considering positive and negative directions, is zero for each dimension. Also, there must be no rotation; thus, the sum of the moments about any point on the body must be zero. In this text we consider only two dimensions, x and y; hence

$$\xrightarrow{+}\sum \mathbf{F}_x = 0 \qquad \textbf{(4.16a)}$$

$$+\uparrow \sum \mathbf{F}_y = 0 \qquad \textbf{(4.16b)}$$

and

$$+\circlearrowright \sum \mathbf{M}_A = 0 \qquad \textbf{(4.16c)}$$

These three equations may be applied to any problem, and an equal number of unknowns may be determined. Should there be more than three unknowns, the system is indeterminate; the analysis of these systems is beyond our present scope.

Example 4.5 For Figure 4.29a, determine the reactions at A and B.

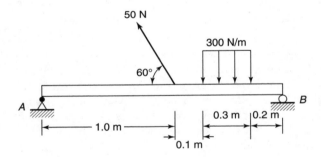

Figure 4.29a

Sketch and Data:
The free-body diagram for the figure is shown in Figure 4.29b.

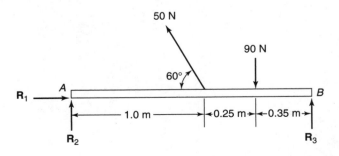

Figure 4.29b

Assumptions:

1. Static equilibrium exists.
2. The container's weight acts at the center of gravity.

Analysis:

The free-body analysis included the two reactive forces at the fixed pin A: a horizontal reactive force to counter the horizontal force in the cable and a vertical reactive force assumed upward. At the roller end B there is only a vertical reactive force upward. Applying the condition of static equilibrium in each dimension yields for the horizontal forces

$$R_1 - 50 \cos 60 = 0$$

$$R_1 = 25 \text{ N} \rightarrow$$

For the vertical force

$$R_2 + R_3 + 50 \sin 60 - 90 = 0$$

$$R_2 + R_3 = 46.7 \text{ N} \uparrow$$

At this point there are two unknowns, so we need a third independent equation, that in static equilibrium the sum of moments about any point is zero. We could pick any point on the beam, but the wisest choice is either A or B, eliminating R_2 or R_3, respectively, from Equation 4.16c. Selecting point A and Equation 4.16b and using the correct sign convention for moments, we get

$$1.6R_3 - (1.25)(90) + (1.0)(50 \sin 60) = 0$$

$$R_3 = 43.2 \text{ N} \uparrow$$

$$R_2 + 43.2 = 46.7 \uparrow$$

$$R_2 = 3.5 \text{ N} \uparrow$$

The assumed vertical direction for the reactive forces R_2 and R_3 is correct. If a negative value for one of the reactions were found, it would indicate that we had assumed the incorrect direction.

Strength of Materials

With the rudimentary understanding of statics and the possible determination of forces acting on an object, we can examine the effect of the forces on the object's material structure. *Stress*, rather than force, is used in analyzing a material's strength and is defined as force per unit area. Stress may be a result of forces acting normal to a surface, normal stresses σ, or acting parallel to a surface, shear stresses τ. We define these as

$$\sigma = F_n/A \qquad (4.17)$$

and

$$\tau = F_t/A \qquad (4.18)$$

where F_n is the normal force, F_t is the tangential force, and A is the cross-sectional area.

Consider the bar shown in Figure 4.30. It is subjected to a force **F**, and because of the force, the bar lengthens a distance ΔL. The ratio of the change in length to the original length is called the *strain* ϵ:

$$\epsilon = \Delta L/L \qquad (4.19)$$

In Figure 4.30 the bar is in tension; the force is pulling the material. In compression, the force pushes on the bar, trying to contract it. The modulus of elasticity E, called *Young's modulus*, is the proportionality constant that relates stress and strain in a material (see Table 4.1). Hooke's law relating stress and strain was discovered by Robert Hooke in the late 1600s:

$$\sigma = E\epsilon \quad \text{or} \quad E = \sigma/\epsilon \qquad (4.20)$$

Equation 4.20 states that stress and strain vary linearly with each other. This is valid for the elastic region of a stress–strain diagram, as in Figure 4.31. If the force is released in the linear (elastic) region, the material will return to its original state. Once the yield point is reached, however, the material does not behave

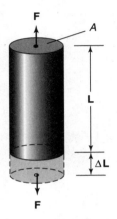

Figure 4.30
A bar subjected to a tensile force.

Table 4.1 Typical values of Young's modulus

Material	Young's modulus E kilopascals (kPa)
Aluminum	7.0×10^7
Brass	9.0×10^7
Steel	2.0×10^8
Glass	7.0×10^7

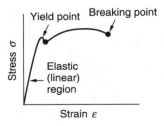

Figure 4.31
A stress–strain diagram.

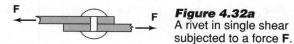

Figure 4.32a
A rivet in single shear subjected to a force **F**.

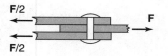

Figure 4.32b
A rivet in double shear subjected to a force **F**.

Table 4.2 Yield stress values for metals

Material	Normal yield stress, MPa	Shear yield stress, MPa
Aluminum	345	207
Bronze	138	76
Steel	620	358

elastically, and a permanent deformation occurs. If stretched far enough, the material will break.

The analysis of materials in tension or compression is similar; the material is acted upon by axial forces. In shear, however, the force is perpendicular to the material's axis. In Figure 4.32a the force acts perpendicular to the cross-sectional area of the rivet. Figure 4.32b shows a rivet in double shear; the shear area is twice the cross-sectional area of the rivet.

When you are designing an object, you must assume a safe allowable value for the stresses and then calculate the actual values, ensuring that the actual values are less than the working value. The normal and shear values σ_y and τ_y, respectively, for materials are known at the yield point. Dividing these values by the desired factor of safety S, yields the allowable value of stress that is used in your sizing calculations (see Table 4.2). Thus,

$$\sigma_{\text{allow}} = \sigma_y/S \tag{4.21}$$

and

$$\tau_{\text{allow}} = \tau_y/S \tag{4.22}$$

For instance, in designing a structural element from aluminum with a factor of safety of 1.75, the allowable normal stress is 197 megapascals (MPa) and the allowable shear stress is 118 MPa.

Example 4.6 A 0.8-m steel bar is machined to two different diameters, 10 and 5 centimeters (cm). The larger diameter is 0.5 m long, and the rod is subjected to a load of 1000 kilonewtons (kN).

Determine the stress in each section and the total extension of the bar under load.

Given:
A bar of two different diameters is subjected to a tensile load.

Find:
The stress in each diameter section and the total elongation of the bar due to load.

Sketch and Data:

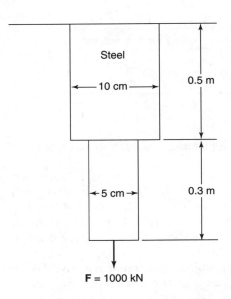

Steel

← 10 cm →

0.5 m

← 5 cm →

0.3 m

F = 1000 kN

Figure 4.33

Assumption:
The bar is in static equilibrium.

Analysis:
From Equation 4.21 we can determine the stress.

$$\sigma_{0.1} = F_n/A = \frac{1000 \text{kN}}{\pi/[4(0.1)^2]\text{m}} = 127{,}324 \text{ kPa}$$

$$\sigma_{0.05} = F_n/A = \frac{1000 \text{kN}}{\pi/[4(0.05)^2]\text{m}} = 509{,}296 \text{ kPa}$$

Since the stresses in each section of the bar are less than the normal yield stress, we know that the bar is in the elastic region and Hooke's law applies. From Equation 4.20, determine the strain and from that the length increase for each diameter section of the bar from Equation 4.19.

$$\epsilon = \sigma/E$$

$$\epsilon_{0.1} = (127{,}324)(2 \times 10^8) = 6.4 \times 10^{-4}$$

$$\epsilon_{0.05} = (509{,}296)(2 \times 10^8) = 2.55 \times 10^{-3}$$

$$\Delta L = \epsilon/L$$

$$\Delta L_{0.1} = (0.5)(6.4 \times 10^{-4})(1000 \text{ mm/m}) = 0.32 \text{ mm}$$

$$\Delta L_{0.05} = (0.3)(2.55 \times 10^{-3})(1000 \text{ mm/m}) = 0.765 \text{ mm}$$

Thus, the total length increase is 1.085 mm, not much at all.

Example 4.7 An aluminum rivet in single shear must withstand a force of 10,000 N. The factor of safety is 1.75. Determine the diameter of the rivet.

Sketch and Data:

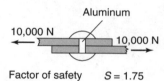

Aluminum

10,000 N ← → 10,000 N

Factor of safety $S = 1.75$ **Figure 4.34**

Assumption:
No deformation occurs.

Analysis:
First determine the allowable shear stress for the rivet, using data from Table 4.2.

$$\tau_{\text{allow}} = 207 \text{ MPa}/1.75 = 118.3 \text{ MPa}$$

$$\tau = F/A = 118\,300 \text{ kPa} = \frac{10 \text{ kN}}{A \text{ m}^2}$$

$$A = 10/118\,300 = 8.45 \times 10^{-5} \text{m}^2$$

$$A = \pi D^2/4$$

$$D = 0.010\text{m} = 1 \text{ cm}$$

A design challenge was made to a group of students to construct a device that featured at least two mechanical subsystems. One student designed and constructed a toggle switch which includes two lever subsystems.

Mechanics Design Report

TOGGLE SWITCH DESIGN

John Buhse
Introduction to Engineering
Section 02

Introduction

The design challenge was to create a quick-disconnect mechanical toggle switch. The switch must remain in the on or off position, much as electric toggle switches do. In this case, the focus is on the mechanical design. There were several questions to address before investigating the design, such as how large the switch should be, what the application will be, and what materials are available to use. It was decided to make the switch and its container 4 to 6 inches in length, so others could readily examine the switching mechanism. It is also easier to construct a larger switch. The machine shop has some extra Lucite, so that became the material of choice for construction of the switch box.

Research and Investigation

An investigation of how switches work was conducted at two levels: one literary, by searching in *The Way Things Work,* and the other forensic, by taking apart an electric toggle switch and seeing the mechanics of the device. From this investigation, several ideas came to mind through brainstorming, such as a slide switch, a screw switch, and a spring-and-cam switch. The slide switch was eliminated because it did not meet the specifications of a quick disconnect; rather, the area of contact gradually decreases, and if it were conducting electricity, this could cause material deterioration due to high current loads. The screw switch was an interesting idea, but it did not meet the desirable goal of having a quick disconnect; rather, turning the switch is required, a more time-consuming operation. The choice of a spring-and-cam switch met the requirements of quick disconnect with full surface contact, but it would be more complicated to design and construct. In this design, the toggle lever is connected to a shaft which rotates a cam through approximately 45°. The cam is in contact with a metal strip, copper, that is fastened at one end to the switch container and is free to move at the other end. The end where it is free to move is where the contacts are located; the cam can displace the strip, opening the contacts, or with the other rotation, allow the strip to move back to engage with the contacts. Figure 4.35 illustrates the final design.

Construction

The construction of the switch box and switch was challenging. The concept of the cam and lever seemed simple, but attaching the various elements together was more complex. The ideal material for the metal strip would be thin and flexible, so when the cam rotated over the material, it would deflect with minimal force. It should be flexible, so it will not deform with repeated use and cease to connect or disconnect the contact as required. Spring steel might be such a material, but it was not available, and copper strips were available in about 1/16-in. thickness. This material was a bit stiff, hence the force required to move the switch from one position to the other increased from 1 to about 12 newtons (N) as the lever moved from one side to the other. The copper deformed slightly with repeated testing, so a return spring was added at the contact end, increasing the force required by 1 N. The shape of the cam was determined by the criterion for quick opening and closing, which translates to a cam with a high rise rate; this shape also increases the moment arm, in effect increasing the disadvantage, which is why the final force was as high as it was. One of the preliminary designs had a rounded cam, but that shape did not displace the lever decisively. In Appendix A the analysis of the toggle lever arm is shown. There were some machining problems with making the cam profile smooth; in addition, the profile does cause the switch to open about 1 cm, hence the force required increases dramatically with displacement. For instance, to cause the initial separation of the contacts requires 1 N, the additional force is for displacing the strip far enough for the cam to rotate to the off position.

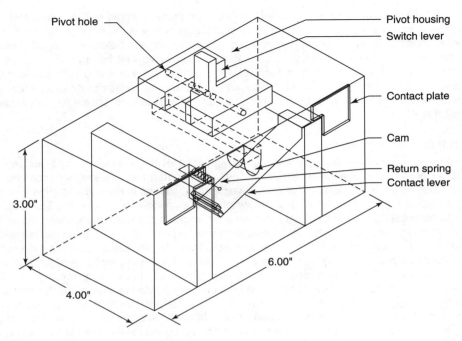

Figure 4.35
Toggle switch.

Testing

The switch meets the design criteria—it is a toggle switch that opens quickly and positively. Testing was done on the lever system to determine the force required to displace the contacts a given distance of separation. Table 4.3 contains the data. The face of the cam was shaped during the testing to make the rise less rapid, reducing the force required for switching.

Table 4.3

Contact separation, mm	Force required, N
5	1
10	1.1
25	1.75
50	3.5
75	6.5
100	12

Conclusion

The cam-and-spring switch fulfills the design requirements that the switch open and close quickly and use a combination of simple lever systems. However, the force required can be a hindrance at times. The force can be decreased in future designs by using a more flexible metal strip, shaping the cam profile so it opens the cam only the desired amount and not farther, and cuts the cam so that the surface is smooth.

Bibliography

1. *The Ways Things Work, An Illustrated Encyclopedia of Technology.* New York: Simon & Schuster, 1967.
2. Burghardt, M. David. *Introduction to Engineering Design and Problem Solving.* New York: McGraw-Hill, 1999.

Appendix

The force balance for the lever arm is shown in Figure 4.36.

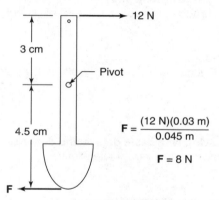

$$F = \frac{(12 \text{ N})(0.03 \text{ m})}{0.045 \text{ m}}$$

$$F = 8 \text{ N}$$

Figure 4.36

Assessment Critique

The toggle switch design is a well-executed project, and the assessment rubrics identify exactly why this is so. What follows is an annotated assessment sheet for the project.

DESIGN ASSESSMENT RUBRICS

The Design Process

A. Identified problem criteria, constraints, and specifications.   0   1   2   ③
Clearly explained the problem with specifications of quick closing and material availability.

B. Gathered background information from a variety of sources.   0   1   ②   3
Good indication of sources for material; did not include information learned from encyclopedia or forensic analysis of electric switch.

C. Suggested several alternative solutions.   0   1   2   ③
Several alternative switches described in sufficient detail so as to know why they were not adequate for the design.

D. Evaluated ideas against design criteria and made improvements.   0   1   2   ③
Improvements made such as adding the spring because of copper deformation.

E. Justified the chosen solution.   0   1   ②   3
This was the only solution that disconnected quickly; alternatives did not, so little justification was required.

The Design Solution

A. Provided an accurate drawing with basic details and dimensions.   0   1   2   ③
Computer-aided design drawing with overall dimensions.

B. Constructed the model and used materials appropriately.   0   1   2   ③
Quality of finished model is very good, visually appealing.

C. The solution worked. It fulfilled the design criteria.   0   1   2   ③
The solution functioned well; quick disconnect and connect, and multiple levers.

D. Originality and creativity of the design.   0   1   ②   3
Good creativity, but the forces required are high.

Testing

A. Used knowledge gained from testing to inform design.　　0　1　2　③
Used testing information to reshape cam.

Work Habits

A. Completed assigned task in a timely fashion.　　0　1　2　③
Project and report turned in on due date.

Communication and Presentation

A. Demonstrated understanding of key ideas orally and/or in writing.　　0　1　②　3
The idea of moments is clearly presented, as are considerations in making switches. The explanation of multiple-lever systems could be amplified.

B. Report neatly written with good grammar.　　0　1　2　③
Report neatly written with no spelling errors or syntax mistakes.

Scoring guide:　　0 = No response or unacceptable response
1 = Acceptable response
2 = Good response
3 = Excellent response

Score: 35
Total possible points: 39

4.34. In Figure 4.21*a,* let force **A** = 100 N, **B** = 80 N, and **C** = 50 N. **Problems**
Determine the resultant force's magnitude and direction.

4.35. It is possible to resolve concurrent forces graphically. Perform the
graphical solution for the situation in Problem 4.34.

4.36. Determine the resultant force (magnitude and direction) for the concurrent force systems in Figure P4.36*a* to *d*.

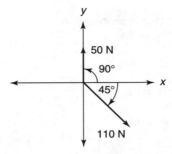

Figure P4.36a

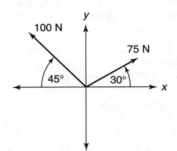

Figure P4.36b

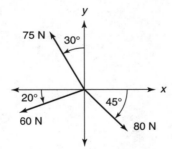

Figure P4.36c

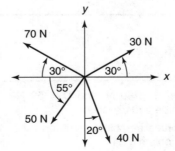

Figure P4.36d

4.37. Two tugboats are pulling a barge as illustrated in Figure P4.37. The horizontal force acting at C is 20,000 N. Determine the force (tension) in ropes (lines in nautical jargon) AC and BC.

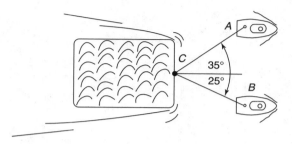

Figure P4.37

4.38. Given that the resultant force acting on the tent peg shown in Figure 4.22a is 1000 N and the angles are as indicated, determine the values of **A** and **B**.

4.39. It is desired to determine the drag force on a boat hull. A model of the hull is placed in a water channel, and water flows past it, modeling a given hull speed. There are lines to prevent the boat from leaving the centerline of the channel as well as a line to pull the boat, with scales to measure the force (tension) in the lines. The readings indicate a tension of 120 N in line AB and 180 N in line AD. Determine the drag force on the hull and the tension in line AC. See Figure P4.39.

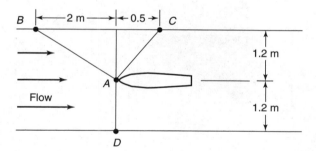

Figure P4.39

4.40. Two connected cables support a load as shown in Figure P4.40. Determine the tension in AC and BC.

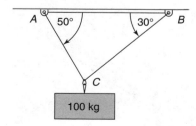

Figure P4.40

4.41. Two connected cables support a load as shown in Figure P4.41. Determine the tension in *AC* and *BC*.

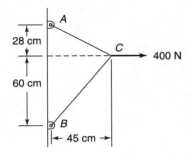

Figure P4.41

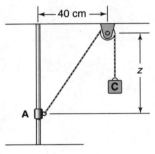

Figure P4.42

4.42. In Figure P4.42, the 6-kg collar **A**, may slide on the frictionless vertical rod. It is connected via a pulley to a 6.8-kg counterweight **C**. Determine the value of height z for which the system is in equilibrium.

4.43. A container and its contents weigh 1000 kN. See Figure P4.43. Determine the shortest possible sling *ACB* which may be used to lift the loaded container if the tension in the sling cannot exceed 750 kN.

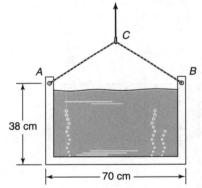

Figure P4.43

4.44. Determine the moment (torque) in foot-pounds about point *A* for the wrench in Figure P4.44.

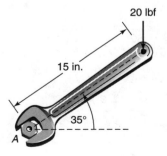

Figure P4.44

4.45. Determine the net moments about A and B for the beam in Figure P4.45.

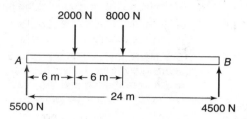

Figure P4.45

4.46. The wheelbarrow must be supported at an angle of 20° to the horizontal in Figure P4.46. Determine the force required to do so.

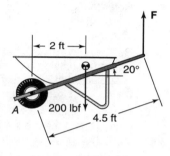

Figure P4.46

4.47. Determine the moment about A for the sketch in Figure P4.47.

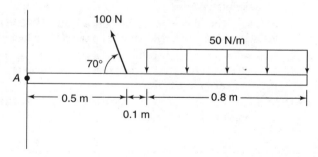

Figure P4.47

4.48. Find the components of the container's weight parallel and perpendicular to the inclined plane in Figure P4.48.

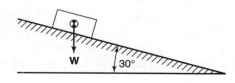

Figure P4.48

4.49. A horizontal beam is 15 m long and weighs 2000 N. It has pinned supports at the extreme left end and 3 m from the right end. In addition, there is a concentrated load of 500 N a distance of 5 m from the left end. Determine the reactions at the supports.

4.50. Refer to Figure 4.29*a* in Example 4.5. Let the angle be 120°, and determine the reactions at *A* and *B*.

4.51. A 15-ft plank that weighs 7 pounds per foot (lb/ft) is horizontal, attached rigidly to the wall at the right end, and supported by a vertical cable 3 ft from the left end. A person weighing 150 lb is standing in the center of the plank. Determine the reactions at both end supports.

4.52. The person in Problem 4.51 now moves past the cable and is standing 1.5 ft from the left-hand side. Determine the reactions at both supports.

4.53. A tire has a diameter of 0.65 m and supports a vertical load of 500 kg. A force, acting at the centerline, must be sufficient to cause the tire to move over a 10-cm curb. What is the amount of the force?

4.54. Refer to Figure 4.30. Let the force be 5000 N, the length 60 cm, and the diameter 2 cm. Calculate the normal stress and strain for steel, aluminum, and brass.

4.55. Refer to Figure 4.30. Let the bar be hollow with an inside diameter of 10 cm, an outside diameter of 12.5 cm, and a length of 1 m. The bar is subjected to a loading of 2000 kN. Determine the elongation and the normal stress if the material is aluminum.

4.56. Refer to Figure 4.33. Let the bar be bored to an inside diameter of 3 cm. Determine the loadings as in Example 4.6.

4.57. Two 1-cm-diameter rivets join two metal sheets. If the force pulling the sheets is 35 kN, determine the average shear stress in each rivet.

4.58. Two plastic parts are butted together and glued. The parts' mating surfaces have dimensions of 100 millimeters (mm) by 5 mm. If a tensile force of 5000 N is applied, what is the average normal stress at the interface?

4.59. A riveted connection must support a load of 9000 N in single shear. What diameter steel rivet should be used if the factor of safety is 1.5?

4.60. A riveted connection with two rivets in double shear supports a 1500-kN load. What is the rivet diameter if the rivets are steel and the factor of safety is 1.5?

4.61. The cable in Problem 4.51 has an allowable stress of 58,500 kPa. Determine its diameter for the worst load condition.

**Electro-
mechanical
Devices**

There are several everyday electromechanical devices such as thermostats, toasters, relays, and solenoid valves that have simple mechanical and electric circuits.

Thermostats

A thermostat is a switch that opens and closes as a function of temperature. The most common type of thermostat, used in electrical appliances such as toasters, is a bimetallic strip. Figure 4.37*a* illustrates such a thermostat, which consists of two strips

Mechanical engineering laboratory where students are examining a linear displacement transformer, a device that converts a linear displacement to an electrical signal. (*Courtesy of Hofstra University*)

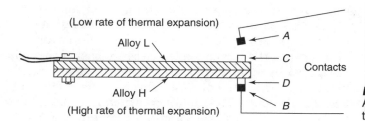

Figure 4.37a
A bimetallic thermostat in the cold position.

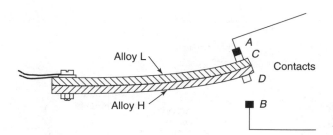

Figure 4.37b
A bimetallic thermostat in the heated position.

of metal bonded to each other. One of the metals has a high coefficient of thermal expansion and the other a low coefficient. At a cool temperature the strip is straight; as the temperature rises, alloy H increases in length compared to alloy L, causing the strip to bend, as shown in Figure 4.37*b*.

The strip has two contacts *C* and *D*, which connect with terminals *A* and *B*. When the temperature is cool, the strip is straight and contacts *B* and *D* connect. As the temperature rises, this connection is broken and at a high temperature *A* and *C* connect. In most applications, only one set of contacts is used; so if the thermostat is used to break a circuit when the temperature is hot, then only contacts *D* and *B* are used. Thus, when the temperature falls, the bimetallic strip straightens, contact is made, and perhaps a heater turns on. The temperature rises, the contacts separate, and the circuit is opened, turning off the heater. In the alternative situation, a thermostat may be used for cooling, for example, air conditioning, in which case contacts *A* and *C* close when the temperature is too hot, causing a cooling process to start. It is possible to vary the temperature at which heating occurs by physically moving the position of the *B* contact closer to or farther from *D* by using a knob or lever, such as a toaster light-to-dark temperature control, which moves the contact.

Toasters

Toasters use a variety of control systems to turn off the circuit when the interior temperature rises to a predetermined level. One such system uses a bimetallic strip and employs simple series and parallel circuits. When the toaster lever is pushed down, it closes an electric switch, the main switch in this case, and engages the spring-loaded carriage basket in a clip that can be opened by a bimetallic strip. When the main switch is closed in Figure 4.38, current flows through the main and auxiliary heating elements. The auxiliary switch is bimetallic, and as the temperature rises, it opens and short-circuits the auxiliary heater resistance. Electricity follows the path of least resistance, so flowing through the switch is easier than through the auxiliary

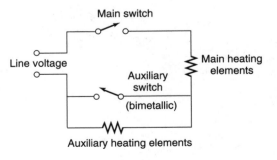

Figure 4.38
Electric circuit for a toaster.

heater. Current is still flowing through the main heating element, and the bimetallic strip begins to straighten; however, it is initially prevented from doing so by a device attached to the carriage basket clip. Eventually the force of the bimetallic strip causes the clip to move, releasing the spring-loaded basket, which moves up, opens the main switch, and releases the bimetallic strip.

A note of caution if you decide to investigate your toaster: Disconnect it prior to any examination. The heating elements, nichrome wire, are not electrically insulated, and it is dangerous to touch them when the toaster is plugged in, even if the toaster is turned off. Figure 4.39 illustrates why this may be so. When the toaster is plugged into the wall, one side of the heating element is directly connected to the outlet. Should this be the hot side of the circuit, if you touch the nichrome wire ($+115$ V) and ground (0 V), a voltage potential exists across you and current will flow through you. Figure 4.40 illustrates a typical grounded household electric outlet. One side is connected to voltage source supply, the hot side, and the other is connected voltage source return, or ground. This means that a voltage potential exists between the hot side of a line and anything else that is physically connected to the earth—hopefully not you! The ground connection on the outlet allows the

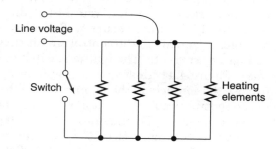

Figure 4.39
Toaster heating elements may be connected to the hot side of the outlet, creating the potential for a shock if a knife or fork is inserted.

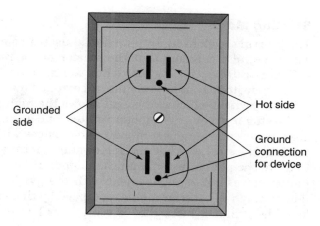

Figure 4.40
Wall receptacle with ground connection.

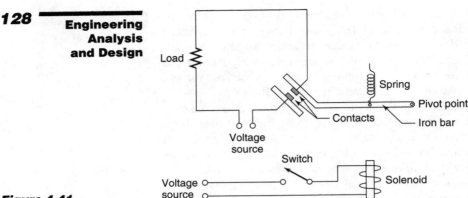

appliance to be grounded when the appliance is plugged into the outlet.

Relays

A relay is used to open and close a switch in another circuit which has high values of current or voltage across it. Relays can be used for safety reasons so that you do not directly engage a switch which has high values of current flowing through it; or there may be circuit elements in the control circuit that cannot tolerate high voltages or currents. The ignition switch in an automobile uses a relay. Figure 4.41 illustrates a simple relay that uses a solenoid to close the contacts in the main circuit. When the switch is closed in the control circuit, current flows through the solenoid, creating an electromagnet. The magnet pulls on the iron bar, closing the contacts in the main circuit. When the switch is opened, the solenoid no longer creates a magnetic field and the iron bar is moved by the spring, opening the main circuit.

Starting Motor

The starting system in an automobile uses a solenoid switch to carry the high currents that the starter motor requires. Figure 4.42 illustrates such a system. When you turn the key and engage the key switch, a small amount of current flows from it to the solenoid switch, which closes and allows the high flow of current to the starting motor. The motor rotates, and a small gear (pinion gear) moves down the motor shaft and engages with the flywheel, rotating the engine. When the engine starts, it causes the flywheel to drive the pinion faster than the motor's rotational speed. The pinion moves out of engagement with the flywheel, and current flow through the starter motor is stopped by the person using the key. The grinding noise that is sometimes heard when a person

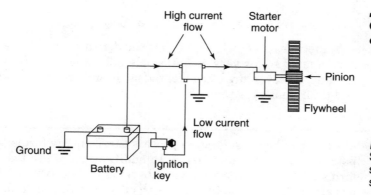

Figure 4.42
Simplified automobile
starting circuit using a
solenoid.

attempts to start an engine that is already running is due to the
pinion hitting the flywheel and not engaging.

Energy and Energy Analysis

Thermodynamics is the science that is devoted to understanding
energy in all its forms (mechanical, electrical, chemical) and how
energy changes form—the transformation of chemical energy
to thermal energy, for instance. It provides the tools necessary
to understand how heat can be converted to work and what
limitations are imposed on that conversion process.

Conservation of Mass

The expression for mass flow rate is frequently used in conjunction
with thermodynamic analysis. Let the mass flow rate in kilograms
per second (kg/s) be denoted as $\dot{m}$. The conservation of mass states
that for steady-state conditions, the mass flow entering a device
must equal the mass flow leaving the device. Thus,

$$\dot{m}_{\text{in}} = \dot{m}_{\text{out}} \tag{4.23}$$

Figure 4.43 represents a pipe with a fluid flowing steadily through
it. At plane 1 there is a velocity v, a density ρ, and an area A.
The same variables exist at plane 2. All three terms may vary,
depending on the substance and the flow conditions. The mass
flow rate at any plane is

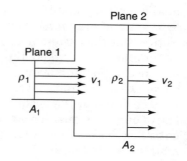

Figure 4.43
One-dimensional fluid flow
in a pipe.

$$\dot{m} = \rho A v = \rho_1 A_1 v_1 = \rho_2 A_2 v_2 = \text{constant} \qquad \textbf{(4.24)}$$

Sometimes students forget that the density can change; this is particularly so for very compressible substances, such as gases. Liquids, however, are often treated as incompressible substances, and their density is then constant.

Energy Forms

Matter may be considered to possess three energy forms; kinetic, potential, and internal energy. In addition, there are two energy forms that may enter or leave a thermodynamic system. What is a system? A *system* is the device or substance that is undergoing, performing, or receiving the energy transformation. Fortunately there are only two types of systems, one for constant mass and one for mass flow through an object. The former system is called a *closed system*; it is closed to mass flow, and it would be the system used if we were to heat 2 kg of water. The 2 kg of water remains constant throughout the heating process. On the other hand, an *open system* allows mass flow through it, such as an automotive engine that has air and fuel entering it and exhaust leaving it. In this case the mass is not constant, hence a different system is required for the analysis.

Work

Work is a force $\mathbf{F}$ acting through a distance Δx, or

$$\mathbf{W} = (\mathbf{F} \text{ N})(\Delta x \text{ m}) \qquad \textbf{(4.25)}$$

where the force might be the horizontal force required to push a wheelbarrow a distance Δx. The N and m refer to the units associated with force (newtons) and distance (meters); thus work has units of newton-meters (N·m), or joules (J). The force can take many forms; it can be the force acting on a mass to raise it, or it can be the force necessary to move a charged particle in a magnetic field. It may be a pressure acting on area, such as a piston crown, causing it to move. When work is performed in thermodynamics, a system is involved. Either the system is performing work on the surroundings (everything external to the system), or the surroundings are doing work on the system. To mathematically distinguish the two cases, we refer to work done *by* a system as *positive* and work done *to* a system as *negative*. Notice that the system does not possess work; work results from an interaction between the system and the surroundings, for instance, a piston moving. The energy that allows the system to do work comes from energy contained by the matter within or passing through the system. Thus, work is an energy form that exists only in transition across a system's boundary. Once work enters or leaves the system, its effect causes a change in the matter's energy form.

Heat

Heat, represented by the symbol Q, is similar to work in that it is not an energy form that matter possesses; *heat* is defined as energy crossing a system's boundary because of a temperature difference between the system and the surroundings. This definition differs from the colloquial use of heat, in that matter does not have an energy form called *heat*. There are situations in which there is no heat flow, although a temperature difference exists between the system and surroundings. We called this an *adiabatic* process. For instance, if a hot water pipe is well insulated from the surroundings, then any heat flow will be very small, negligible in most cases, and the pipe is modeled as an adiabatic, open system. The units of heat are joules, the same as those of work; but the sign convention for heat is opposite to that for work: Heat flow into the system is positive, heat flow from the system is negative.

Potential, Kinetic, and Internal Energies

For many applications in engineering, matter may be viewed as containing three energy forms: gravitational potential, kinetic, and internal energies.

The potential energy of a system mass depends on its position in the gravitational force field. There must be a reference datum and a distance z from the datum to the system mass. If a force **F** raises the system, the change of potential energy PE is equal to the work (force through a distance) necessary to move the system.

$$\Delta(\text{PE}) = \mathbf{F} \, \Delta z = (m \text{ kg})(g \text{ m/s}^2)(\Delta z \text{ m}) = mg \, \Delta z \quad J \quad (4.26)$$

where the gravitational acceleration is assumed constant over the distance the mass moves.

The kinetic energy of a system is developed in an analogous manner. Again let us consider a system of mass m. A horizontal force acts on the system and moves it a distance x. Since it moves horizontally, there is no change in potential energy. The change in kinetic energy (KE) is defined as the work required to move the system a distance Δx.

$$\text{KE} = \mathbf{F} \, \Delta x$$

$$\mathbf{F} = m \, \mathbf{a} = m \, \frac{\Delta v}{\Delta t}$$

where

$$\frac{\Delta v}{\Delta t} = \frac{\Delta x}{\Delta t} \frac{\Delta v}{\Delta x} = v \frac{\Delta v}{\Delta x}$$

and where the average velocity v is defined as

$$v = \frac{v_1 + v_2}{2}$$

Because $\Delta v = v_2 - v_1$, the change in kinetic energy becomes

Table 4.4 Specific heats for
various substances

Substance	c kJ/kg · K
Air	0.7176
Aluminum	0.963
Brick	0.92
Bronze	0.4353
Concrete	0.653
Gasoline	2.093
Glass	0.833
Ice	1.988
Steel	0.419
Water (liquid)	4.186
Water (vapor)	1.403
Wood	2.51

$$\Delta(\text{KE}) = \frac{m}{2}(v_2 - v_1)(v_2 + v_1) = (\frac{m}{2}\text{kg})(v_2^2 - v_1^2 \, m^2/s^2) \quad J$$

(4.27)

Note that this is the change of translational kinetic energy of the system; if the velocity of the system increases, the kinetic energy increases. If a car is moving at 40 kilometers per hour (km/h) and is accelerated (force through a distance) to 80 km/h, the kinetic energy increases. If the car's velocity is zero, its kinetic energy is zero.

One of the less tangible forms of energy of a substance is its *internal* energy. This is the energy associated with the substance's molecular structure. Although we cannot measure internal energy, we can measure changes in internal energy. The symbol for specific internal energy is u and for total internal energy U.

$$u = \text{specific internal energy, J/kg} \qquad \textbf{(4.28a)}$$

$$U = mu, \text{ total internal energy, J} \qquad \textbf{(4.28b)}$$

The change in internal energy for a substance may be written as

$$\Delta U = (m \text{ kg}) [(c \text{ kJ/kg} \cdot \text{K})](T_2 - T_1 \text{ K}) \qquad \text{kJ} \quad \textbf{(4.29)}$$

where c is a property called *specific heat*. Typical values for specific heat are given in Table 4.4.

**The
First Law of
Thermodynamics**

The conservation of energy, or first law of thermodynamics for a constant-mass system, may be developed as follows. Assume a system of constant mass has an initial amount of energy, receives energy in the form of heat, produces some work (energy out), and has a final amount of energy left. Let $E = U + \text{KE} + \text{PE}$ represent the total energy of the substance. The conservation of energy is

Energy initial $\quad+\quad$ Energy added

$\qquad E_1 \qquad\qquad + \qquad\quad Q$

$\qquad\qquad\qquad\qquad = \quad$ Energy final $\quad+\quad$ Energy produced

$\qquad\qquad\qquad\qquad = \qquad\quad E_2 \qquad\quad + \qquad\qquad W$

$$(4.30)$$

This may be combined to form

$$Q = E_2 - E_1 + W \qquad (4.31)$$

If this equation is divided by the mass m, the quantities have units of energy per unit mass (kJ/kg), which is denoted by the use of lowercase letters:

$$q = e_2 - e_1 + w \qquad (4.32)$$

Expanding the total energy term into its components yields

$$Q = m(u_2 - u_1) + \frac{m(v_2^2 - v_1^2)}{2} + mg(z_2 - z_1) + W \quad (4.33)$$

As you use the first-law equations, remember that the sign convention for heat and work must be used when substituting the numeric value in the equation.

Example 4.8 An adiabatic pump casing holds 5 kg of water and receives 2 kW of power to the water as it moves through the pump. The discharge valve of the pump is inadvertently closed, and the pump delivers the power to the water in the casing for 1 min before the valve is reopened. What is the water's temperature rise after 1 min?

Given:
An adiabatic pump contains a known mass of water and receives work for a certain period.

Find:
The water's temperature rise due to the work.

Sketch and Data:

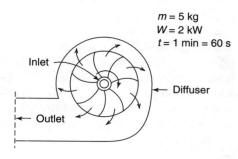

$m = 5$ kg
$W = 2$ kW
$t = 1$ min $= 60$ s

Inlet

Diffuser

Outlet

Figure 4.44

Assumptions:
1. The water in the pump casing is a closed system.

2. The heat is zero (adiabatic), and the changes in kinetic and potential energies are zero.

Analysis:

The first law for a closed system is

$$Q = \Delta U + \Delta KE + \Delta PE + W$$

Applying the assumptions and writing the expression for the change of internal energy in terms of temperature yield, where $W = \dot{W}t$,

$$0 = mc(T_2 - T_1) + (-5 \text{ kJ/s})(60 \text{ s})$$

$$0 = (5 \text{ kg})[4.186 \text{ kJ/kg} \cdot \text{K}](\Delta T \text{ K}) - 300 \text{ kJ}$$

$$\Delta T = 300/5 \cdot 4.186 = 14.3 \text{ K or } °\text{C}$$

The conservation of energy for open systems, the first law for open systems, may be viewed similarly to the closed system; this time the system receives a heat flux and produces power while the fluid flows through it. The energy entering the control volume is the fluid's energy plus any heat flow; the energy leaving the control volume is the fluid's energy plus any work done by the fluid within the control volume. The first law may be expressed as

$$\text{Energy in} = \text{Energy out}$$

$$\dot{Q} + \dot{m}_{in}(e + p/\rho)_{in} = \dot{W} + \dot{m}_{out}(e + p/\rho)_{out} \qquad \textbf{(4.34)}$$

where the heat flux is

$$\dot{Q} = (m \text{ kg/s})(q \text{ kJ/kg}) \qquad \text{W} \qquad \textbf{(4.35)}$$

and the power is

$$\dot{W} = (m \text{ kg/s})(w \text{ kJ/kg}) \qquad \text{W} \qquad \textbf{(4.36)}$$

For liquids, the p/ρ term may be neglected, but not for gases which are compressible.

Example 4.9   A fluid enters a device with a steady flow of 3.7 kg/s, an initial pressure of 700 kPa, an initial density of 3.2 kg/m³, an initial velocity of 50 m/s, and an initial specific internal energy of 2000 kJ/kg. It leaves at 150 kPa with a density of 0.6 kg/m³, a velocity of 150 m/s, and a final specific internal energy of 1950 kJ/kg. The heat loss from the device is 18 kJ/kg. Determine the power into or out of the device.

Given:

An open system receives a fluid with known properties, heat and work interactions occur, and the fluid leaves with known properties.

Find:

The power.

Sketch and Data

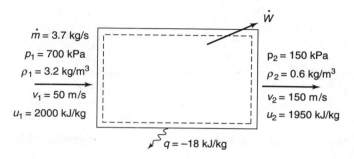

$\dot{W}$

$\dot{m} = 3.7$ kg/s
$p_1 = 700$ kPa
$\rho_1 = 3.2$ kg/m³
$v_1 = 50$ m/s
$u_1 = 2000$ kJ/kg

$p_2 = 150$ kPa
$\rho_2 = 0.6$ kg/m³
$v_2 = 150$ m/s
$u_2 = 1950$ kJ/kg

$q = -18$ kJ/kg

Figure 4.45

Assumptions:

1. It is an open steady-state system.
2. Neglect changes in potential energy.

Analysis

The first law for open systems is

$$\dot{Q} + \dot{m}_{in}(e + p/\rho)_{in} = \dot{W} + \dot{m}_{out}(e + p/\rho)_{out}$$

$$e_1 = 2000 \text{ kJ/kg} + \frac{50^2 \text{ m}^2/\text{s}^2}{2(1000 \text{ J/kJ})} = 2001.25 \text{ kJ/kg}$$

$$e_2 = 1950 + \frac{150^2}{2(1000)} = 1961.25 \text{ kJ/kg}$$

$$\dot{Q} = (3.7 \text{ kg/s})(-18 \text{ kJ/kg}) = -66.6 \text{ kW}$$

Substituting into the first-law equation yields

$$-66.6 + 3.7(2001.25 + 700/3.2) = \dot{W} + 3.7(1961.25 + 150/0.6)$$

$$\dot{W} = -34.22 \text{ kW or power into the device}$$

A power plant contains many of the fundamental elements that thermodynamic analysis is involved with—heat exchangers, turbines, and pumps. Figure 4.46 illustrates a simplified power plant. The heat is typically supplied by burning fuel, but other heat sources include solar energy and nuclear energy; the heat boils water in a steam generator. The high-pressure, high-temperature steam flows to the turbine, which rotates, producing power. The steam exits the turbine at a low pressure and low temperature to a condenser, a heat exchanger, where it is condensed to a liquid and finally pumped back to the steam generator. Some of the power produced by the turbine is visualized as being used to

Energy Analysis

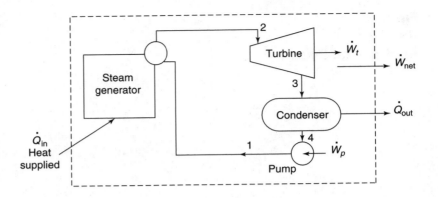

Figure 4.46
A simplified power plant.

power the pump, hence there is net work produced by the steam power plant.

The heat added $\dot{Q}_{in}$ must be equal to the energy leaving as net work $\dot{W}_{net}$, the heat leaving the condenser $\dot{Q}_{out}$. The heat out is negative and the heat in is positive, and if they are algebraically added, they yield the net work. This is one of the fundamental laws governing power-producing cycles.

$$\dot{W}_{net} = \dot{Q}_{in} + \dot{Q}_{out} \qquad (4.37)$$

If the heat added is from the combustion of fuels,

$$\dot{Q}_{in} = \dot{m}_f h_{RP} \qquad (4.38)$$

where $\dot{m}_f$ is the fuel flow rate in kg/s and h_{RP} is the heating value of the fuel, the heat released per unit mass by the combustion process. Table 4.5 lists some typical heating values for commonly used fuels.

An ideal power plant's efficiency, how well it converts heat to work, can be described as

$$\eta_{th} = 1 - \frac{T_C}{T_H} = \frac{\dot{W}_{net}}{\dot{Q}_{in}} \qquad (4.39)$$

where T_C is the cycle low temperature, the condensing temperature, and T_H is the cycle high temperature, the combustion temperature, while η_{th} is the thermal efficiency. This efficiency in terms of temperature in Equation 4.39 is valid for a Carnot cycle, but its definition in terms of work output divided by heat input is valid in general.

Table 4.5 Average heating values for various fuels

Fuel	kJ/kg
Coal	27,900
Corn cobs (dry)	21,600
Gasoline	44,800
Lignite	26,500
Natural gas	57,450
Residual oil	43,000
Wood (dry)	20,350

Example 4.10 A power plant uses residual oil with a density of 990 kg/m³ as its fuel. The plant produces 1000 MW of net power and is characterized by a high temperature of 1000 K and a low temperature of 300 K. Determine the amount of fuel required for one week's supply and the size of two cylindrical tanks, $L = D$, to hold the fuel.

Given:
A power plant with the net power and characteristic temperatures and fuel.

Find:
The fuel supply required and tank size to hold the fuel.

Sketch and Data:

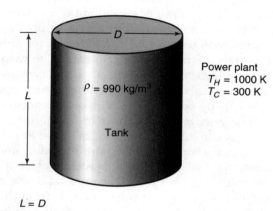

$\rho = 990 \text{ kg/m}^3$

Tank

$L = D$

Power plant
$T_H = 1000 \text{ K}$
$T_C = 300 \text{ K}$

Figure 4.47

Assumption:
The power plant follows the theoretical model.

Analysis:
The thermal efficiency may be used to find the heat required and from that the fuel flow required.

$$\eta_{th} = 1 - 300/1000 = 0.7$$

$$0.7 = \dot{W}_{net}/\dot{Q}_{in} = 1000/\dot{Q}_{in}$$

$$\dot{Q}_{in} = 1000/0.7 = 1428.6 \text{ MW}$$

The heat supplied is also equal to mass flow rate times the heating value of the fuel:

$$1,428,600 \text{ kW} = (\dot{m}_f \text{ kg/s})(h_{RP} \text{ kJ/kg}) = (\dot{m}_f)(43,000)$$

$$\dot{m}_f = 33.22 \text{ kg/s}$$

and the weekly fuel required is

$$\dot{m}_{week} = (33.22 \text{ kg/s})(3600 \text{ s/h})(24 \text{ h/day})(7 \text{ days/wk})$$

$$= 20\,093\,425 \text{ kg/wk}$$

The volume that the fuel occupies is the mass divided by the density:

$$V_f = 20\,093\,425 \text{ kg/990 kg/m}^3 = 20\,296 \text{ m}^3$$

Thus each tank must hold 10,148 m³ and

$$10{,}148 = \frac{\pi D^2 L}{4} \quad \text{and} \quad L = D$$

so

$$D = 23.46 \text{ m}$$

Very often residual fuel and coal have constituents, such as sulfur, that create environmental impacts when burned. In the combustion process, sulfur forms sulfur dioxide, which in combination with water, say, rainwater, forms sulfuric acid. It is one of the main ingredients of acid rain. The carbon dioxide formed in the combustion process hastens the greenhouse effect, or warming of the earth. One can calculate the magnitude of these terms quite easily and begin to realize the enormity of the problem.

The gravimetric analysis of fuel tells us the percentage of its various constituents on a mass basis. For instance, the fuel may contain carbon, hydrogen, sulfur, water, and ash. Let us assume that the fuel contains 1 percent sulfur and that the cleaning processes (scrubbers) on the exhaust stack of the steam generator remove 98 percent of sulfur dioxide from the combustion process. Calculate the kilograms of sulfur dioxide produced daily by the power plant in Example 4.10.

The total mass of sulfur emitted daily is

$$\dot{m}_S = 0.01 \dot{m}_f = (0.01)(2\,870\,208) = 28\,702 \text{ kg/day}$$

However, 98 percent of this is contained in the stack, so only 2 percent of this value leaves the stack:

$$(\dot{m}_S)_{\text{emitted}} = (0.02)(28\,702) = 574 \text{ kg/day}$$

The sulfur reacts with oxygen in the air to produce sulfur dioxide according to the following reaction equation:

$$S + O_2 = SO_2$$

$$32 \text{ kg S} + 32 \text{ kg O}_2 = 64 \text{ kg SO}_2$$

The reaction equation is simply a conservation of mass; in this case, 1 mole (mol) of sulfur, 32 kg, reacts with 1 mol of oxygen, 32 kg, to form 1 mol of sulfur dioxide, 64 kg. Thus, for every 32 kg of sulfur burned, 64 kg of sulfur dioxide is formed, or there is a 2-to-1 ratio. Let us just consider the sulfur that escapes, or 1148 kg of sulfur dioxide released. This amounts to 1.26 tons/day from this one plant. Imagine the amount that would be released if the scrubbers were not installed! Not all industries, nor all countries, particularly economically poor countries, use exhaust gas scrubbers.

Hydraulics— Automotive Brakes

Most students are familiar with bicycle hand brakes. You squeeze the brake on the bicycle handle, and a wire pulls calipers together on the wheel's rim, slowing the bike. This is an effective system

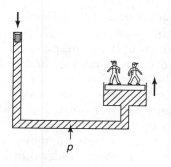

Figure 4.48
The pressure is the same
throughout the hydraulic
system.

for a bicycle, but it would be virtually impossible to stop an automobile by using such a technique. Fortunately, another system is available that is really a hydraulic simple machine. A simple machine is able to use a small force acting over a greater distance to move a larger force acting over a shorter distance.

We know that pressure is force divided by area, $p = F/A$; and if the area becomes small, the pressure, even for a small force, can be very high. For instance, a pin can puncture many hard surfaces with little application of force because the area is small and the pressure is quite high. Figure 4.48 illustrates a simple hydraulic system. The piping is filled with an incompressible liquid such as oil or water, and the pressure throughout the liquid is the same. It is possible to hold two people on a large disk with a comparatively small force if such a force acts on a piston of small cross section. Imagine that the people weigh 200 kg total and are standing on a circular disk 1 m in diameter. The force they exert is

$$F = ma = (200 \text{ kg})(9.8 \text{ m/s}^2) = 1960 \text{ N}$$

The area of the disk is $\pi D^2/4 = \pi(1)^2/4 = 0.7854 \text{ m}^2$; thus the pressure it creates is

$$p = F/A = 1960/0.7854 = 2495 \text{ N/m}^2 = 2495 \text{ Pa}$$

If the small piston diameter is 2 cm, then its area is $\pi(0.02)^2/4 = 0.0003142 \text{ m}^2$, and the force needed to act on this area to create 2495 Pa is $F = pA = (2495)(0.003142) = 0.7841 \text{ N}$, a very small force. Thus, a small force may balance a large force in a hydraulic system in a similar manner to pulleys used to reduce the force needed to lift a heavy object or an inclined plane to push a heavy object. If we want to raise the people 1 cm, the work is

$$W = \int F \, dx = F \, \Delta x = (1960 \text{ N})(0.01 \text{ m}) = 19.6 \text{ N·m} = 19.6 \text{ J}$$

For the smaller piston to create this work, the distance moved is much greater:

$$W = 19.6 = \int F \, dx = (0.7841 \text{ N})(\Delta x \text{ m})$$

$$\Delta x = 25.0 \text{ m}$$

a tremendous distance! This can also be visualized by thinking of the volume of liquid that must move the larger piston upward, and that volume is created by the movement of the small piston.

In hydraulic braking systems, the distance that the brake pads must move is very small, and the area of the piston is comparatively small also, perhaps an inch or less in diameter. The movement of the brake pedal, a lever, is transmitted to a piston in the hydraulic brake system, which moves, creating a pressure increase. The pressure increase causes a piston to move, acting on the wheel brakes. In cars, there is a master cylinder that contains the brake fluid reservoir and transmits the pressure; and most frequently with power-assisted brakes, the pedal movement actuates a pump that increases the pressure in proportion to the pedal pressure. On most cars, the front brakes and some rear brakes are disk brakes, as illustrated in Figure 4.49; the spinning disk is attached to the wheel and the brake pads, pushed by the brake fluid, squeezing it. The rear brakes are often drum brakes (Figure 4.50); a rotating drum is attached to the wheel, and a hydraulic cylinder with two pistons is connected to the brake system. The pressure causes the pistons to move outward, forcing the brake pads to act against the spinning drum.

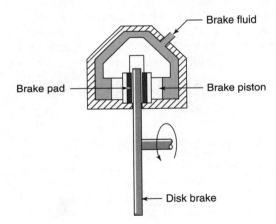

Figure 4.49
Disk brake.

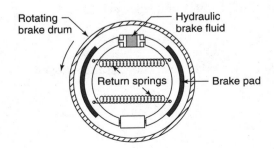

Figure 4.50
Drum brake.

What causes lift on a wing? What causes a sailboat to move in the direction of the wind? The answers to these questions lie in understanding the first law of thermodynamics in its fluid dynamics rendition, known as *Bernoulli's equation*. The first law for open systems is

$$\dot{Q} + \dot{m}(e + p/\rho) = \dot{W} + \dot{m}(e + p/\rho)_2$$

and if we consider a case in which there is no heat transfer ($Q = 0$), no work ($W = 0$), the density is constant, and the temperature is constant ($\Delta u = 0$), then the first law becomes

$$\frac{v_1^2}{2} + g z_1 + (p/\rho)_1 = \frac{v_2^2}{2} + g z_2 + (p/\rho)_2$$

$$\frac{\Delta p}{\rho} + \frac{(v_2^2 - v_1^2)}{2} + g(z_2 - z_1) = 0 \qquad \textbf{(4.40)}$$

which is Bernoulli's equation.

Let us examine air flow over a wing, illustrated in Figure 4.51. As the air approaches the wing, it must divide and recombine after passing over the wing. The distance that the air travels across the top is greater than that traveled by the air below the wing because of the curved surface. In examining Bernoulli's equation for constant elevation ($\Delta z = 0$), we note that if the velocity increases across the top, the pressure must decrease; thus, the pressure is greater on the underside of the wing, and lift occurs. The purpose of the airplane's engine is to drive the plane through the air so that there is an air velocity across the wings and lift can occur.

The sail on a sailboat operates on the same principle when the boat is heading towards the wind (the boat cannot sail directly into the wind). The wind travels a greater distance across the outside of the sail than across the inside, creating a pressure difference across the sail, as illustrated in Figure 4.52. A component of the pressure is in the forward direction; note that pressure times area is force, which acts to push the boat ahead. Further assisting the boat is the keel, which resists sideward movement and redirects a portion of this force ahead. This is called

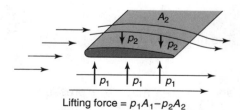

Lifting force = $p_1 A_1 - p_2 A_2$

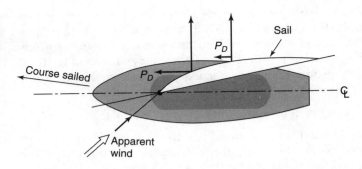

Figure 4.52
Wind flow past a sail.
The pressure drop across
the sail creates a net
driving pressure P_D. This
pressure acting on the sail
area is the driving force,
propelling the boat.

pointing into the wind; and the better a boat points into the wind,
the shorter will be its path in the windward direction without
requiring tacking.

Windmills and Drag

A windmill is a marvelous invention, extracting power from the
wind's energy and providing useful mechanical power to us. Today
wind farms are under development to produce electric power by
using the wind's energy, a source of nonpolluting energy supply.
Because wind velocity fluctuations do not normally coincide with
power requirements, most wind power systems include energy
storage systems. There is a theoretical maximum amount of
power that can be extracted from the wind, as the next example
illustrates.

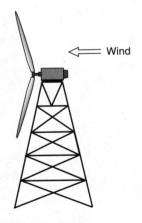

Figure 4.53
A two-bladed horizontal
windmill.

Example 4.11 A windmill has a blade diameter D and receives air with a velocity v and density ρ. The exit velocity from the
blades can be considered negligible. Determine an expression for
the maximum power the windmill can develop. Figure 4.53 illustrates
the windmill.
  The first law for an open system is

$$\dot{Q} + \dot{m}(u + \text{ke} + \text{pe} + p/\rho)_{\text{in}} = \dot{W} + \dot{m}(u + \text{ke} + \text{pe} + p/\rho)_{\text{out}}$$

There is no heat transfer from the windmill, the change of potential
energy is zero, as is the change in pressure divided by density, and
the temperature of the air is constant, hence $\Delta u = 0$. Taking all these
matters into account yields

$$\dot{W} = \frac{\dot{m}v^2}{2}$$

The mass flow rate is

$$\dot{m} = \rho A v$$

and the area may be expressed in terms of the diameter as $A = \pi D^2/4$, so the power becomes

$$\dot{W} = \frac{\pi \rho D^2 v^3}{8}$$

In actual windmills, there is an air exit velocity from the blading, which reduces the theoretical maximum value to 59.3 percent of the value determined above. The windmill power is a function of the velocity cubed, so the windmill is very sensitive to velocity changes.

When an object (e.g., car, boat, airplane) moves through a fluid, power is required to move it a certain distance (force times distance divided by time) and to overcome frictional resistance between the fluid and the object, most often called *drag*. The fluids have viscous properties, which are a measurement of its internal friction and which act to retard motion through the fluid. When a stream of air, for instance, flows past a sail, the particles of air immediately adjacent to the fabric are brought to rest. These particles do not travel along the sail with the main airstream, but instead remain fixed relative to the sail. As a result of the air's viscosity, the stationary particles exert a braking action on their immediate neighbors, which are slowed somewhat. These particles, in turn, slow down the particles next to them, and the process repeats so that it is only at some distance from the sail surface that the layers of air reach their full speed. This is illustrated in Figure 4.54.

The region between the sail and the mainstream in which the particle velocities are less than the stream velocity is termed the *boundary layer*. As the air particles in the boundary layer are slowed down, they lose kinetic energy. This loss in energy is transmitted as friction forces through the air layers until finally it is communicated to the sail as friction forces acting parallel to the sail surface. Thus, the frictional forces are proportional to the velocity squared (kinetic energy loss), and the power required to overcome these forces is proportional to the velocity cubed (force times velocity). The study of boundary layers is an important area of fluid mechanics, as the interaction between the object and fluid

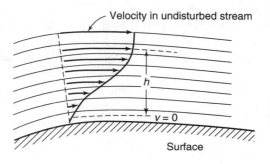

Figure 4.54
Velocity distribution across a boundary layer where *h* is the boundary layer thickness.

determines the boundary layer, which, in part, determines the drag forces acting on the object. Note that the power extracted from the wind varies as the velocity cubed, so it is not surprising that the frictional power loss would vary in the same way. Consider driving in a car at 50 mi/h and then accelerating to 70 mi/h. This represents a 40 percent increase in velocity, but a 174 percent increase in velocity cubed and an equally dramatic increase in drag.

Thermal Engineering Design Report

In this section, students were asked to develop a design project that required an understanding of the laws of conservation of mass and energy. The following design project on food dehydration focuses on designs that will reduce the mass of a moist food product.

Problems

4.62. A steam turbine has an inlet steam flow of 4 kg/s with a density of 20 kg/m³. The inlet diameter is 10 cm, and the outlet diameter is 20 cm. The outlet density is 10 kg/m³. Determine the inlet and outlet velocities.

4.63. Water with a density of 1000 kg/m³ flows steadily through a pipe with an internal diameter of 5 cm. The volume flow rate is 0.5 m³/s. Determine the mass flow and velocity.

4.64. Two gaseous steams containing the same fluid enter a mixing chamber and leave as a single stream. For the first gas, the entrance conditions are $A_1 = 500$ cm², $v_1 = 130$ m/s, and $\rho_i = 1.60$ kg/m³. For the second gas the entrance conditions are $A_2 = 400$ cm², $m_2 = 8.84$ kg/s, and $\rho_2 = 1.992$ kg/m³. The exit stream conditions are $v_3 = 130$ m/s and $\rho_3 = 2.288$ kg/m³. Determine the total mass flow leaving the chamber and the velocity of the second gas entering the chamber.

4.65. One hundred kilojoules per kilogram (kJ/kg) is added to 10 kg of a fluid while 25 kJ/kg of work is extracted. Determine the change in internal energy. Find the temperature change if the substance is (*a*) water and (*b*) air.

4.66. A container holds 15 liters (L) of water. If 2000 kJ of heat is added, what is the temperature change?

4.67. An adiabatic tank contains 2 kg of water at 20°C and receives 20 kN · m of work from a paddle wheel. Determine the final temperature.

4.68. Two kilograms of boiling water (100°C) is poured into a 0.7-kg steel container at 20°C. What will the final equilibrium temperature be, assuming no losses to the surroundings?

4.69. Determine the energy release from burning (*a*) 1 metric ton of coal, (*b*) 1 L of gasoline (specific gravity = 0.836), and (*c*) 5 kg of natural gas.

Problems continued on p. 151

DESIGN OF A SOLAR DEHYDRATOR

Nancy Forsberg
Introduction to Engineering
Section 02

Introduction

The design challenge is to create a solar dehydrator for drying vegetables or fruit. After some investigation, I decided to limit the drying to apples and bananas; these are firm fruits that have a strong skeletal structure, as contrasted to tomatoes, for instance. The materials for constructing the dehydrator were foamboard, Mylar, metal screening, Plexiglas, and fasteners. Since ancient times, food has been preserved through various methods of dehydration and/or drying. Solar drying and salting are two methods to remove water from foods. Solar drying is the method used in this design. There are many advantages to dehydrating food. By removing the water from a food, space and weight are saved in the transportation, holding, and packaging of food, and spoilage should be reduced, too.

Research and Investigation

There are commercial operations that use solar dehydration in which factories have retractable roofs that open to the sun while air, propelled by fans, flows over the dehydrating food. The risks of microbial (disease-causing bacteria) spoilage are reduced while the quality of the food, along with the vitamins and minerals, is preserved. The microbial population requires water for reproduction and growth, and removing it delays food spoilage.

The terms *dried* and *dehydrated* can be used interchangeably; however, dehydration is really the correct term to use in this case. Dehydration refers to the process of drying in a dehydrator. Solar dehydration is the use of the sun's energy to heat the air that is used to remove the water from the food. There must be an airflow over the food to help keep the moisture from settling to the bottom of the food being dried. The airflow can occur naturally due to density variations in the air caused by heating or induced by a fan. The process of dehydration involves simultaneous mass and heat transfer, as energy must be provided to the water in food to evaporate the water. The process is often viewed as an adiabatic one in that energy just evaporates the water and does not heat the remaining food structure above ambient conditions.

Construction

In thinking about dehydration, the importance of airflow led me to want to incorporate a fan mechanism of some kind into the design. Based on pictures of industrial models, I selected three different rectangular devices, all lined with Mylar paper. The first would be an open box approximately 5 in. high and 11 in. square. The fruit was supported by 0.5-in. wire mesh that was suspended 2.5 in. from the bottom. Air was gently blown over the fruit, bottom to top, to prevent moisture accumulation on the bottom of fruit slices. The same-size base was then covered with a pyramid Plexiglas top, in order to trap the sun's energy. Vent holes were drilled in the Plexiglas to allow the air to escape. Last, the base was covered with a Mylar top to which the sides were open. In the latter design, very little direct sunlight would hit the fruit.

To discover which of these three designs for the dehydration of food is most effective, all three were tested simultaneously. Initial testing indicated that little dehydration occurs without airflow, so a 6-in.-diameter fan, used for room ventilation, was employed. The three boxes were placed in the sun, with a fan attached to ductwork that gives them equal airflow. The airflow from the fan was directed to an 8-in.-diameter duct made of rolled up construction paper (oaktag) with one end blocked. This became the main air supply, and branch lines—paper towel tubes with 0.25-in.-diameter holes to disperse the air across the bottom of the dehydrator—fed the air to each of the dehydrators. The airflow was tested with an air velocity indicator, and the paper towel diameter was crimped to balance the flow.

Analysis and Testing

Into each of the dehydrators, 50 g of thinly sliced banana and 50 g of thinly sliced apple were placed. The weight of the fruit and wire rack was measured at various times in the dehydration process and recorded. The weight of the rack, 56 g, was subtracted from each of the measurements.

The process was done twice to ensure that the results were reproducible. Both sets of results yielded the same conclusion. The bananas and apples dried more completely in the dehydrator that had no roof and was open to the air. It had been anticipated that the Plexiglas roof dehydrator would be the most effective, but it was not. Upon reflection, it seems as if without a roof, the air can leave the dehydrator most easily, and there may be more air movement caused by naturally occurring breezes. Figure 4.55a and b is a plot of the data. Figure 4.56 illustrates the solar open-top dehydrator.

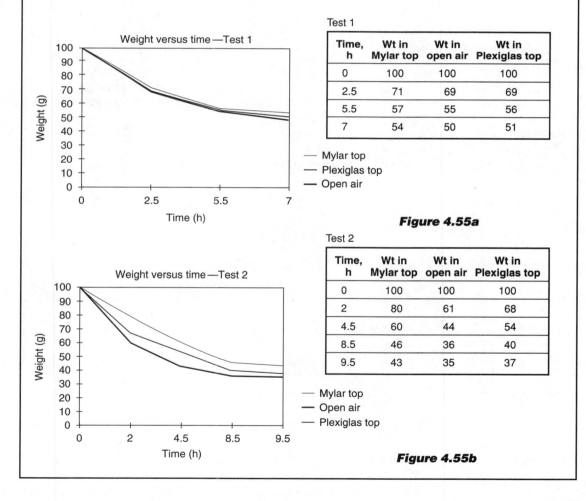

Test 1

Time, h	Wt in Mylar top	Wt in open air	Wt in Plexiglas top
0	100	100	100
2.5	71	69	69
5.5	57	55	56
7	54	50	51

— Mylar top
— Plexiglas top
— Open air

Figure 4.55a

Test 2

Time, h	Wt in Mylar top	Wt in open air	Wt in Plexiglas top
0	100	100	100
2	80	61	68
4.5	60	44	54
8.5	46	36	40
9.5	43	35	37

— Mylar top
— Open air
— Plexiglas top

Figure 4.55b

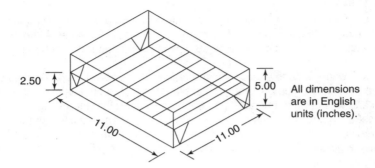

2.50

5.00

All dimensions
are in English
units (inches).

11.00

11.00

Figure 4.56

Conclusions

The least expensive design proved to be the most effective design for the dehydration of apples and bananas. It is not known whether the 5-in. height is a variable. It may be that lower sides would make a more efficient design, so several alternative dehydrators of different heights should be tested to discover the optimum height. In examining the drying curves, a good deal of the drying occurs because of only airflow, as the Mylar-top dehydrator had significant weight reduction.

References

Borgstrom, Georg. *Principles of Food Science*, vol. 1. NY: Macmillan, 1968.
Singh, R. P.; and D. R. Heldman. *Introduction to Food Engineering.* New York: Academic, 1991.

Assessment Critique

The Design of a Solar Dehydrator is a well-executed project, and the assessment rubrics identify exactly why this is so. What follows is an annotated assessment sheet for the project.

DESIGN ASSESSMENT RUBRICS

The Design Process

A. Identified problem criteria, constraints, and specifications.   0   1   2   **(3)**
Good discussion of constraints on design and specifications in terms of dehydration.

B. Gathered background information from a variety of sources.   0   1   **(2)**   3
Note your sources in the discussion. Expand slightly in doing so.

C. Suggested several alternative solutions.   0   1   2   **(3)**
Several alternative designs were discussed, and they will be tested and evaluated.

D. Evaluated ideas against design criteria and made improvements.   0   1   **(2)**   3
Did not explain why you selected the designs you did for testing; noted they came to mind.

E. Justified the chosen solution.   0   1   2   **(3)**
The testing is used objectively to select the best design.

The Design Solution

A. Provided an accurate drawing with basic details and dimensions.   0   1   2   **(3)**
Provided computer-aided design drawings with overall dimensions and details.

B. Constructed the model and used materials appropriately.   0   1   2   **(3)**
Observation of the models showed the quality of workmanship and appropriate use.

C. The solution worked. It fulfilled the design criteria.   0   1   2   **(3)**
Dehydration occurred in all designs; in some better than in others, as expected.

D. Originality and creativity of the design.   0   1   **(2)**   3
The designs are good, solid, and straightforward.

Testing

A. Used knowledge gained from testing to inform design.   0   1   2   **(3)**
Testing is a key element of this project; knowledge from testing indicates best design.

Work Habits

A. Completed assigned task in a timely fashion.   0   1   2   ③
The project and design report were turned in on time.

Communication and Presentation

A. Demonstrated understanding of key ideas orally and/or in writing.   0   1   2   ③
The oral presentation was fine, with good explanation of testing procedure.

B. Report neatly written with good grammar.   0   1   2   ③
Report well written, with spelling, syntax, and structure fine.

Scoring guide: 0 = No response or unacceptable response
1 = Acceptable response
2 = Good response
3 = Excellent response

Score: 36
Total possible points: 39

4.70. A tank with a diameter of 2 m and a height of 3 m is located 50 m above the ground. It is filled with water from a pump located on the ground. What energy is required to fill the tank? If the tank fills in 1 hr, what average power is required?

4.71. A 0.5-kg container is dropped from the top of a 75-m-tall building. Determine its kinetic energy and potential energy

(a) At the moment it is dropped

(b) After it has fallen 50 m

(c) The instant before it hits the ground

4.72. An airplane weighing 10,000 kg is flying 2000 m above the earth's surface at 1000 km/h. Determine the plane's kinetic and potential energies.

4.73. A rifle bullet has a mass of 1.5 g and leaves the barrel of the gun at 500 m/s. Determine its kinetic energy.

4.74. Five people are lifted on an elevator through a distance of 100 m. The work is found to be 343 kJ. Determine the average mass per person.

4.75. An adiabatically insulated 2-kg container is dropped from a balloon 3.5 km above the earth. Upon impact with the ground, the box remains intact; the volume remains the same, so no work is done on it. What is the change in internal energy in the box after impact?

4.76. A fluid enters a device with a steady flow of 3.7 kg/s, an initial pressure of 690 kPa, an initial density of 3.2 kg/m^3, an initial velocity of 60 m/s, and an initial specific internal energy of 2000 kJ/kg. It leaves at 172 kPa, $\rho_2 = 0.64$ kg/m^3, and $u = 1950$ kJ/kg. The heat loss is found to be 18.6 kJ/kg. Find the power (work per unit time) in kilowatts.

4.77. A fluid at 700 kPa, with a specific volume of 0.25 m^3/kg and a velocity of 175 m/s, enters a device. Heat loss from the device by radiation is 23 kJ/kg. The work done by the fluid is 465 kJ/kg. The fluid exits at 136 kPa, 0.94 m^3/kg, and 335 m/s. Determine the change in internal energy.

4.78. An air compressor handles 8.5 m^3/min of air with a density of 1.26 kg/m^3 and a pressure of 1 atmosphere (atm) and discharges it at 550 kPa with a density of 4.86 kg/m^3. The change in the specific internal energy across the compressor is 82 kJ/kg, and the heat loss by cooling is 24 kJ/kg. Neglecting changes in kinetic and potential energies, find the power in kilowatts.

4.79. Calculate the kinetic energy of a 1200-kg automobile moving at 60 mi/h.

4.80. The automobile in Problem 4.79 is stopped. The brakes have an average specific heat of 0.92 kJ/(kg · K). Assume that one-half of the energy is adiabatically absorbed by the brakes, which have a collective mass of 6 kg. Determine the temperature rise of the brakes.

4.81. A heat power cycle with a thermal efficiency of 0.4 produces 12,000 kJ of net work. Determine the heat added and heat rejected per cycle.

4.82. A heat power cycle with an efficiency of 35 percent receives 1500 megawatts (MW) of heat. Determine the net power produced in MW.

4.83. Refer to Example 4.10. Determine the heat leaving the power plant. If water is used for cooling, receiving the heat from the power plant, and it increases from 15 to 25° C, what is the required flow rate in cubic meters per second?

4.84. The cooling water in Problem 4.83 comes from a lake where the return is mixed. Atmospheric cooling at night maintains a stable temperature. However, the specifications require that the lake be large enough such that the mixing of the 25° C water into the lake water at 15° C will not cause the lake water to increase in temperature more than 0.5° C in a 24-h period. How large a volume must the lake have? (The density of water is 1000 kg/m^3.)

4.85. The power plant in Example 4.10 now uses coal with a 3 percent sulfur content. Determine the sulfur dioxide produced, with the same scrubber efficiency mentioned in the text, and the tons of coal required each day. If a railroad car holds 86,000 kg, how many carloads of coal are needed per week?

Computer Applications

In addition to the Internet, which is transforming how the world transmits information and conducts business, the computer has transformed engineering design and manufacturing with the use of software applications. The computer has become an essential tool for engineers. Imagine a circle with smaller circles tangent to it. The inner circle represents a computer, and the outer circles are applications such as word processing, database management, desktop publishing, equation solvers, finite-element analysis, and spreadsheets. Each of these applications is linked to the others through the computer. As computing power increases, the inner circle's diameter increases and more applications and interconnections are possible.

Before the advent of the industrial age, people hand-crafted devices; with the industrial revolution, people operated machines which made the devices; and in the computer era, people supervise a computer which controls a machine which makes the devices. These machines range from automated teller machines (ATMs) at banks, to industrial robots, to numerically controlled milling machines. Notice that the supervisor's knowledge must expand and include an understanding of how the computer operates as well as the machine. Engineers use a variety of software packages to supervise the computer's operation and need to know how information is transmitted between programs and within programs.

The most common universal code is the American Standard Code for Information Interchange (ASCII), a 7-bit alphanumeric code that allows 128 (2^7) code combinations for letters, numbers, and symbols. Basically there is an ASCII code for each of the characters on a keyboard; note that uppercase and lowercase letters have different ASCII codes, for instance, S = 1010011 while s = 1110011. Data files in one program may be converted to ASCII files and read by another program.

Computer Programming

When digital computers were first developed in the 1940s, and for about 10 years following, people had to communicate with the computer in machine language. Whereas this language is handy for computers, it is decidedly not easy to program and hence is error-prone. Every computer manufacturer had its own machine language, and the computer instructions and memory addresses were specified in digits. To add two values might include these instructions:

```
150 58 410034
151 53 410038
152 50 421050
```

Machine languages, known as first-generation programming languages, were too difficult to use, and engineers soon developed a more symbolic language, assembly-level programming, known as a second-generation language. In the late 1950s and early 1960s third-generation languages were developed, including BASIC and FORTRAN. In the 1970s fourth-generation languages were created such as Pascal and C, and additional languages were created in the 1980s such as C++ and Ada. Students may take a course in computer programming as part of their undergraduate curriculum, using texts such as *C Program Design for Engineers* by A. Tan and T. D'Orazio or *Introduction to Fortran 90/95* by S. Chapman.

Computer-Aided Design

The acronym CAD stands for *computer-aided design*. It most often appears in conjunction with CAM, computer-aided manufacturing. CAD/CAM refers not to one activity in the design or manufacturing process, but to many activities that are enhanced by using a computer. For complex designs the computer system frequently used is a workstation linked to the mainframe. The workstation, a very powerful computer, enables the engineer to call up programs from the mainframe and use them in creating a design or performing an analysis. The results can be seen locally and communicated back to the mainframe. The same information can be transmitted from the mainframe to computers used in the manufacturing process.

CAD is used in virtually all engineering fields and includes software that assists designing in electrical engineering as well as structural engineering. The word *CAD* at the first-year student level is associated with computer-aided drawing. In this situation a wire-frame model is often created and then refined into one that is more representative of what we see in three dimensions. New software allows engineers to start with three-dimensional drawing, or solid modeling. In this situation, reminiscent of the computer-generated images seen in movies, the part can be modeled with internal and external surfaces and any intermediate parts, all with different colorations. Sophisticated workstations can light the object at different angles, producing very artistic effects, truly a blending of art and engineering.

At this point the part has been designed, analyzed, and redesigned in light of that analysis, and it can be manufactured. The information about the design component, its CAD data file, can be sent to the computer controlling the manufacturing process, perhaps a manufacturing robot or a numerically controlled machine. The data from the CAD system must be compatible with the CAM system, and manufacturers need to invest in automated manufacturing systems that use CAD/CAM in the first place. In traditional engineering practice, a prototype is often made of a device before proceeding to manufacture; however, as the software for modeling has improved and simulates the actual features with excellent accuracy, some companies are moving directly from the computer model to manufacturing. Boeing did this in the manufacture of its 777. There are implications for companies doing business with Boeing or any other company that uses CAD/CAM: In addition to the part or subsystem that is being contracted for, there is a need for the data file describing it which can be used in the design and analysis.

Equation Solvers

In addition to computer programming and perhaps more frequently in lieu of programming, engineers will use equation-solving software, such as Mathcad or MATLAB®. These software packages use sophisticated computer programming and allow engineers to solve complicated equations without having to be encumbered with writing the computer programs. The focus is on correctly modeling the physical situation and making certain that the problem's assumptions are compatible with the assumptions within the software. This is not usually a difficulty for students in engineering courses, but it may pose problems for practicing engineers when they need to know that the limits of the software are compatible with the situation they are analyzing. Two texts often used for Mathcad and MATLAB instruction are *Introduction to MATLAB for Engineers* by W. Palm and *Mathcad: A Tool for Engineering Problem Solving* by P. Pritchard.

Spreadsheets

Spreadsheets are an application that gained initial popularity with the accounting and financial world and now are important in engineering practice as well. Engineers use them for tracking and projecting project costs. These programs analyze data that are located in a precise way within the programs. Two of the more popular programs are Lotus 1-2-3 and Excel. The data analysis involves separating interrelated information into constituent parts, then varying the parts to determine the effect on the whole. The data must be separable and arranged in rows and columns. The array of numbers that is created may be manipulated with various mathematical techniques. The intersection of a row and column is a *cell*. A cell may contain a label (alphanumeric or word symbol), a value (numeric value), or a formula. In most spreadsheets when you start a cell with a letter, the program assumes you are entering a label. Values must start with a number, and a formula with a sign, typically a plus sign, alerting the program that an equation is being entered. The area where the data is entered is called a *worksheet*.

For commercial applications, the cells can number in the thousands, and the analysis can be quite complex. You can appreciate the importance of the underlying equations that manipulate the data, and as engineers, you are expected to know the validity of the equations used, with their associated assumptions. There may be courses that include spreadsheet solutions as well as the use of equation solvers. A useful reference and text for engineering use of spreadsheets is Byron Gottfried's *Spreadsheet Tools for Engineers: Excel 97 Version*.

Word Processing

Word processing is an odd term for software that eases the tedium of writing; note that it does not create words of wisdom, but does allow whatever words are written to be used easily and edited quickly and efficiently. Computers were first created for data manipulation, or processing; later, as more sophisticated software was created, words could be manipulated also, or processed, hence the terminology.

Word processing software is menu-driven and can be used almost immediately, as the screen on which you see the typed copy has indexes to help sessions. Not only can you move quickly through the text and move chunks of text from one location to another, but you can also replace words easily. Many programs have a text replace feature, so a word can be replaced throughout the text, which is useful for correcting a misspelling. The software includes a dictionary, spell checker, and usually a thesaurus, which provides alternative words to make your writing more interesting to the reader. You can import text from one document to another, so retyping the material is not necessary. In addition, you can determine how you would like the material formatted and see it on screen before printing. The WYSIWYG feature (What You See Is What You Get) is common to current word processors.

For students, word processing is very useful in all your courses, from English to engineering. The papers and reports you write have the potential to reflect very well on you—both content and format matter. Engineering does require significant communication, as noted in Chapter 3, so word processing will be an essential part of your job as you prepare reports and proposals. The text by A. Eisenberg entitled *A Beginner's Guide to Technical Communication* is a useful reference for technical report writing.

Databases and Database Processing

Information is the key to making engineering and managerial decisions. Knowledge about new products, sales, manufacturing costs, inventory, parts, and personnel becomes important. Not only are these subjects individually important, but also their interrelationships can be extremely valuable at times. Knowing the in-house inventory of parts may reduce the time required to undertake a new product line. Database technology allows associated data to be processed as a whole.

Consider a university, an environment in which you are immersed at the moment. Different areas of the university need related information each semester. For instance, information about faculty teaching load is required to generate professors' paychecks, while some of this information is needed to schedule who is teaching what, when, and where. Related to the class schedule are the student data regarding who is attending what classes so grades may be given. Figure 4.57 illustrates these three application files and the data that are required in them. Each file contains its own data, even though those same data will appear in more than one file.

Database problems occur when you want information that crosses file boundaries, such as the average salary of faculty teaching a certain course. This requires information from the faculty data file and the class data file. There is no certainty that the information in one file will be formatted in such a way as to be accessible by another file. Sometimes it is simply not worth the effort to get this information from the computer, unless the information is stored in a database (Fig. 4.58).

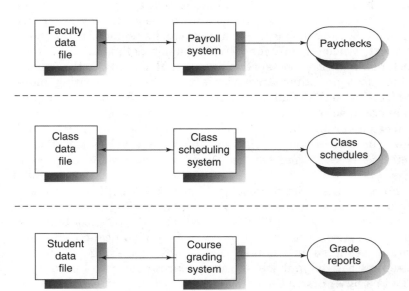

Figure 4.57
Three application files used in universities with no communication between files.

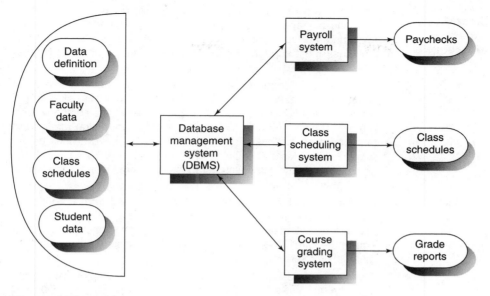

Figure 4.58
An integrated database management system eliminates boundaries between data files and thus allows information sharing.

In this situation a database management system (DBMS) acts as a system librarian. It stores the data and their description and retrieves these data for application programs as needed. The application programs have not changed their function; they are just integrated into a whole by the DBMS. Several advantages spring to mind regarding DBMS: You can obtain more information from a given piece of data, as it may be correlated in a variety of ways; data duplication is reduced or eliminated; and inconsistencies in the data are eliminated, such as student or faculty names. The DBMS is more complex than a file system, and the application programs need to be more sophisticated to handle the generic data. Because DBMS is an integrated system, failure in one part of the system can cause the entire system to be down, and in the case of failure, recovery is more difficult. The primary disadvantage to adopting DBMS is the conversion of data to the database format and revising specific applications programs to accept the data in a more sophisticated form. However, the advantages of greater information gain are driving the industry to adopt DBMS more fully, particularly as it can run on minicomputers as well as mainframes.

The following is a list of possible design projects; many include some fundamental engineering analysis.

Design Problems

4.86. Design and build a full-size chair made from corrugated cardboard and masking tape. The seat must be between 16 and 18 in. from the floor, and the top of the back must be at least 30 in. from the floor. The chair must support at least a 220-lb person and be as light as possible.

4.87. The marketing department at your company has found that videotape sales are increasing and that a product that organizes, stores, and displays the videocassettes will be profitable for the company. You are given the assignment of creating a mock-up design made from foamboard or cardboard along with sketches and drawings of the product. Your design should demonstrate aesthetic as well as functional aspects and may include fabrication techniques for the actual materials.

4.88. Design and construct a vehicle that is suspended from a 40-ft steel wire by screw eyes and that will carry a raw egg as quickly as possible across that distance and not break the egg when impact occurs at the end of the wire. The propulsion system should not use combustible materials for safety reasons.

4.89. A clever gymnastics coach wants to ensure that his gymnasts do not injure themselves when they are practicing and learning new flips and handsprings. Design and build a model of the device that fits around the gymnast's waist as a belt and that would allow for twisting, flipping, and turning motions. The device will prevent the person from falling.

4.90. Design and construct a working model of a 13-in. ruler which only has six marks on it, but can be used to measure an integral length between 1 and 13, that is, 1, 2, 3, 4, ..., 13.

4.91. Design and construct a model of a folding platform such that when extended, the stage is 12 ft wide, 10 ft deep, and 2 ft high. Sections are to be hinged to each other so that the depth after folding is 2 ft and the height 3 ft. The structure is to be mounted on rollers that can be retracted when it is set as a stage.

4.92. Design and construct a model of an overhead garage door system suitable for opening a door 7 ft high and 16 ft wide. The door will be divided into four horizontal sections that are hinged together and supported at the sides by rollers running in a track. Counterbalancing should be included to assist in opening the door.

4.93. Design and construct a device for sorting coins of various denominations—pennies, dimes, nickels, and quarters.

4.94. Design and construct a model of a playground roller coaster in which the starting platform is about 4 ft high and the track is about 12 ft long, including dips and curves. Arrangement should be convenient for passengers to get into the car, which must have the capability of running on the track and then coasting on the ground.

4.95. Design and construct a model of a four-passenger children's merry-go-round that is 10 ft in diameter. The seats are equipped with handles and footrests that oscillate forward and back to propel the passengers in a circular orbit.

5

Discussions with Practicing Engineers

What is life like for engineers working for large companies? For consulting firms? For municipalities? This chapter explores these questions and in the process provides another view of the challenge of engineering. Rich Kumpfbeck is the director of the antenna design laboratory at GEC-Marconi Hazeltine. He has worked at Hazeltine for more than 30 years and witnessed the change, the evolution, of Hazeltine from a high-quality electronics equipment company with a primary, perhaps sole, focus on defense work, to a company that now has defense and commercial business. It has taken a shift in mind-set on the part of engineers to create this vibrant high-technology corporation successful in both arenas.

When the company realized that the defense contracts were not going to be as plentiful in the future as they had been in the past, corporate executives, virtually all engineers, held strategy meetings to analyze their assets—their employees collectively and individually. They found certain areas of expertise and then tried to find a match in the commercial marketplace. Defense industries are challenged with different constraints when working on government contracts, including maximum performance and the highest possible reliability rather than minimum cost. There were other factors as well, but cost is not the paramount issue. In the commercial world, cost is the paramount issue. The products must be cost-competitive; to build a better mousetrap is not sufficient, it must work better than and be as cost-effective as the old-fashioned variety to secure a market.

Experience with the military had given Hazeltine unique expertise that it could apply to the commercial marketplace. Hazeltine's standards of construction are very high to ensure the quality demanded of military components; the manufacturing processes have to provide very small tolerances, hence low variability and

Hazeltine's Rich Kumpfbeck

159

highly consistent products. The defense background also provided the engineers with experience in antennas that operate in hostile environments. Part of their expertise was in creating high-frequency antennas for a wide range of communications, including the frequency bands of cellular phones.

Analysis of the cellular phone system showed that there were two types of antennas: those on millions of phones, which of necessity had to be very inexpensive, and those on the towers that service the cellular base stations. The tower antennas were larger, more complex, and hence more expensive and therefore a possible market. The company's sales force visited cell phone companies to find out what the problems were with the existing antennas. These problems became an opportunity for a new product. The company also purchased existing antennas to see what could be improved, and if there were patents to work around or new ones that could be sought based on new designs.

The cell operators had a variety of problems to address. The antennas had to operate over a wide temperature range; for instance, in the desert, the daytime temperatures soar to 150° F while the night temperatures drop to 40° F, subjecting the antenna and its components to thermal expansion and contraction. If this were not enough, the antennas are on top of towers which vibrate in the wind. Mechanical problems induced by vibration can easily affect the performance of the cellular telephone system. The same antennas also had to be resistant to moisture and, if located near the coast, to corrosion from salty air. The cell operators provided some of the data, and profiles of the vibration were measured. These data were used in the analysis of the new designs so that the designs would allow a long life. Supports were built into the design to damp the vibration and extend the life of joints subject to expansion and contraction. Figure 5.1 illustrates a cellular phone antenna.

Figure 5.1
Internal construction of a cellular phone tower antenna. (*Courtesy of GEC-Marconi Hazeltine.*)

These problems became challenges for the engineers—technical challenges that they knew they could meet and the financial challenge to create the design for the right price. There were performance problems to face as well. One is called *passive intermodulation product performance*. Many times, two or more people will be on the phone at the same time, sending a signal to the antenna and receiving a signal from the antenna. Two or more simultaneous transmission signals may interact and form a third signal, which creates interference and poor reception for the primary signals. Of course, this is to be avoided, and special features are required to minimize this interference.

The switch to the commercial world required a focus on time: the commercial world anticipates a design cycle time of three to four months; the defense industry, three to four years. In three years, Hazeltine built a product line of 12 antennas to provide coverage for a variety of beam widths, the same time period a military contract might specify for the completion of one antenna. Military antenna development is performance- and reliability-driven, not cost-driven, so the constraints on the problem yield solutions in different time scales.

For instance, in the commercial marketplace, an experienced design engineer can perform a quick paper design, many times referred to as a back-of-the-envelope design, which can be used for material and manufacturing cost estimates. If the cost is such that the company can make a profit, then a more formal design is created after a contract is secured. Military defense companies know that they have inherently higher overhead to meet defense specifications in terms of testing and analysis, so their commercial designs must require fewer hours and materials to be cost-competitive with products of companies that produce only commercial products.

The cost is estimated early in the design process to determine whether the company will proceed with a proposal or move into a market. When it is considering a new antenna, the first question Hazeltine must resolve is whether it can design and manufacture the antenna profitably. An experienced engineer develops the design strategy, including time estimates for initial design, construction of a model, testing, and the final redesign. A beginning engineer may be involved in the design of antenna components.

Where does the money come from when one is deciding to build an antenna and enter this marketplace? Management has to devote some of the company's resources to develop the concept and build the prototypes for marketing and sales engineers to sell. In the case of an antenna, it might cost $30,000 in up-front resources for personnel and materials to complete the antenna prototype that will sell for about $700, of which approximately $100 will be profit. The remaining $600 covers the expense of manufacturing, shipping, and marketing the antenna. The market analysis must indicate that a sufficient number of units can be sold so that the company can repay itself the $30,000 as well as allow stockholders to receive dividends and the company to reinvest in new technologies. The demand is not just for one product, but once it is created, to improve upon the product, developing additional lines to complement it.

An engineering notebook, or design journal, is an important part of every engineer's day. This journal is a record of what you are working on, summarizing your experience for the day, and this is particularly important in patent work, where you want to document as early as possible in the design process when an idea was conceived. Many engineers work on computers most of the day, and there are software organizers that provide this function as well, but the engineer has to summarize the information. Sometimes engineers are pulled off projects or projects are terminated because of lack of funding, so the knowledge gained is stored in the notebook. This helps on similar future projects, where the knowledge gained can be revisited.

What does an engineer do? Solve problems. When starting to work at a company, you will learn the company procedures and design software and procedures. The procedures can be reporting procedures—whom do you work with, how do you order supplies? The design tools may be company-specific or more generic and similar to software used at your university. Once the company knows that you have the necessary tools to work on a task, you will be given one. Remember that in designing the antenna the senior engineer knew that a variety of subsystems or components needed detail design; you will work on this detail design such as creating lightning protection for the antenna dipoles. The exact design analysis, size, and specifications are what you will find. The general design approach was decided by the senior project engineer, and you will be part of the team implementing that approach.

As part of the design, you specify certain sizes and values for components and set tolerances for each one. A tolerance analysis provides you with information such that if the values for the components vary within the specified tolerances, the device will still provide acceptable performance. A robust design is one that will work with dimensional variations created by manufacturing and variation in component value. For instance, a 100-Ω resistor may have an actual value of 99 Ω. The variation of 1 Ω in a single resistor may not be significant, but the cumulative effect of property variations of many components can be.

As a beginning engineer you may also be called upon to establish a testing procedure for the antenna; a technician will often run the test and gather the data, but you analyze them and recommend design modifications if the performance is not on target. The analysis and design tasks that you are given are well defined. Aspects that require creative insight are expected after several years of experience, not something necessary from the outset.

As a new engineer, you should be well aware that budget and schedule are the two primary constraints driving the commercial engineering field, where creativity is required to find innovative approaches to solving problems that provide timely, low-cost solutions. Engineers are assisted in this quest by having better analytical tools than 20 or 30 years ago. The computer software support is, in general, excellent, and the machining capability is much more accurate. A note of caution when you are working with software is the issue of accuracy: Does the software provide the correct solution for the conditions you impose? Experience with a particular piece of software gives greater assurance that the result is correct, but sometimes an incorrect result occurs, so checking the answer for reasonableness is important. You will work on computers about 80 to 90 percent of the time, not only for software analysis, but also for sending e-mail to colleagues with questions and writing the necessary documentation in support of your work.

Communicative ability—writing and speaking—is important to your engineering career. There are a variety of reports, such as quarterly status reports, engineering accomplishment reports, and final design reports. The customer will typically specify the report format desired; certainly in defense work, where documentation is a significant requirement, perhaps 10 percent of a project cost will be devoted to documentation. This contrasts with commercial work, where very little documentation is provided, just the product data sheet and perhaps information on product maintenance and installation, but none on the design. However, the commercial world has its own demands for your writing skills—proposal writing. In this situation a customer has a need, a requirement, and solicits proposals on how to meet this need. The company's proposal will have to include a general method of attack and costs, hence the company needs to invest resources to create the proposal. The proposal needs to be very readable, highlighting the positive aspects of your solution and downplaying any potential problems. This is a skill that you hone as you move higher in the organization and one aided from the outset by writing reports and proposals.

Proposals will often have a 30- to 45-day response time, meaning that essentially in six weeks a company must put together a complete and competitive response. A small team of people devotes about two to three weeks to deciding how to solve the problem and developing a preliminary design procedure with cost estimates. Another two weeks is spent in writing the first draft of the proposal. Often at this point, another team of engineers from the company, not associated with the project, critiques the proposal, and a final draft is written based on their input.

Concurrent engineering is practiced at most companies, although its meaning will vary from company to company. At Hazeltine, weekly meetings are held for all projects, with members from marketing, manufacturing, engineering, and quality control attending. At the start of and throughout the project, it is important for all these engineers to communicate their concerns and insights about the design. The weekly meetings bring everyone together to make sure the project schedule and procedures remain compatible for all parties. In some situations, a team from all disciplines is formed, and offices are temporarily relocated, to work on project with a tight deadline. In these situations there may be overtime pay, typically not something engineers receive as salaried professionals, as there will be an extended workweek.

Many, if not most, high-technology companies provide a dual ladder system of promotion. All engineers begin at the same point, but some are more interested in management and less interested in design and may move to that ladder, supervising people rather than creating products. The dual ladder system attempts to create pay equity so that creative people will stay in product design and development and not switch to management, where the salary scale has been historically higher. Technical sales may be viewed as another ladder, although marketing is often viewed as part of the management ladder. You will find out which path most appeals to you, recognizing your aptitudes, interests, and abilities as you gain work experience.

Study the product designs around you. Ask yourself why they are designed as they are. Can they be improved? This will help develop your abilities as a design engineer, and typically people are very happy to discuss something they created. These can make for interesting lunch and coffee-break conversations. Similarly, if you are interested in management, what are the attributes of successful managers in your company? What are the paths they followed? How can you develop these abilities within the corporate organization?

Rich was asked, Why should someone study engineering? The response was immediate: Engineering is fun, challenging; no problem is ever the same, never boring or repetitive. If you like solving problems, stay technical regarding the dual ladder system, as management typically deals with people problems, not creative solutions to new technical challenges. It is very rewarding to see your designs in manufacture, to see the products you designed, or contributed to, awaiting shipment to customers. Engineering is also a terrific background for other fields, as nowhere else are you educated to be disciplined, creative, innovative, and able to define and solve problems.

Engineers often entertain the idea of creating their own companies. One avenue for achieving this is by being a sales representative. For instance, a company in California may need someone to represent its products in New York and visit interested customers. This representative must have the technical skills to intelligently discuss the various products with fellow engineers and be able to call upon others in the main office in response to detailed questions. It is possible to represent several companies at once, typically in noncompeting markets. Perhaps a more frequent aspiration is to create a company that makes a product rather than providing a service. Rich Kumpfbeck relates that about 15 years ago, he and a friend, also an engineer, had a new product idea for a police radar detector. They both worked for Hazeltine, which did not make the product, nor did it want to; so there was no conflict of interest or potential patent infringement. They decided to start their own business on a part-time basis and see if it could evolve into a new company. After seven years of success and near success (at one point they had 40 full- and part-time employees manufacturing and selling the detectors), they went out of business. Why?

The problems are ones that have occurred many times and are typically ones that cause many new businesses to fail, even though the businesses have a better mousetrap at a competitive price. Not only do you need a great idea, but also you must be driven to be successful. Rich and his colleague worked their full-time jobs and then came home to work four or five hours more, plus weekends. They became stretched too thin and needed to devote full-time effort to the project. They did not fully realize how important the marketing of a product is and the need for full-time marketing support. Their initial capitalization allowed for manufacturing but not for the level of sales and marketing personnel actually required. Perhaps in doing it over again, either Rich or his colleague would work full-time for the start-up company along with full-time marketing support. This increases the risk if the business does not succeed, but positively provides the commitment and energy necessary to launch a new business. Profits must be reinvested, improving (innovating) existing products and creating new ones to expand the product line. This is difficult to do and can generate some conflict when you also want to reap some of the rewards for the hard work in creating the venture.

When one is selling in the retail commercial world, there are a variety of costs, sometimes viewed as parasitic, but ones that exist nonetheless. As an example, for a radar detector that retails for more than $200, the manufacturer might make it for $80 and add $40 to cover marketing and profit; the manufacturer sends it to a distributor, who in turn adds $40 for expenses and profit; and finally the retailer adds the final $40+ to the product for expenses and profit. There can be a frustration, even resentment, that others are getting rich from your great ideas, but those ideas have to be distributed and sold. Direct marketing will bypass some of these expenses, but will create other problems to contend with.

Business Plan

If you listen to people who have created their own businesses, or read about them, you may find the term *business plan* discussed. This term describes the essential process that an entrepreneur goes through before launching a business venture and is really a compilation of several plans, each addressing different aspects in manufacturing a product. The product plan describes what you will be selling, how much research and development is required (cost), and how the product line will develop over time. If this is to be an ongoing business, then there will be additional products to complement the initial design. You are aware of the importance of marketing, and the marketing plan describes what the market is—why customers will buy your product and how much they will spend for it, including reasons for their being receptive to your new idea. It also identifies your competition and projects what sales will be as a function of time. The marketing plan identifies the potential cash flow into the business, which is absolutely essential in obtaining start-up capital. The manufacturing plan illustrates how the product can be made, and answers the questions of what facilities are required, whether equipment will be purchased or the task outsourced, what it will cost to make it. Who will be doing what in the organization? The personnel plan resolves that issue as to what jobs need to be filled by whom. As the business grows, more personnel will be needed and should be identified. Where is the money coming from? The finance plan addresses the capital requirements—how much, from whom, and when. There may be progressive financing based on the company's performance, so certain achievement milestones are identified. Investors in your business will read this section with intense interest. Finally, there is a cash flow spreadsheet, indicating the monthly cash flow for the next year with estimates for later years.

MTA's Jerry Burstein

Jerry Burstein, currently working for the Metropolitan Transit Authority (MTA) in quality assurance, has been involved with testing, manufacturability, and quality control with a variety of companies for 30 years. In this role, he seeks designs that are compatible with periodic testing, helping to ensure operational reliability. Customers realize a device has a certain lifetime, it will eventually fail and need to be replaced; but the failure should be diagnosable, and this requires testing. You will want to isolate a circuit and check for defective components, so there should be access to these components for testing. It may be wise to have self-testing features added to integrated circuits, minimizing the need for external test points.

Jerry started his career in engineering during the Apollo mission to the moon, a time of engineering triumph in the late 1960s. The triumph was short-lived; soon after the successful moon missions, government funding of R&D was significantly reduced by the early 1970s. Government funding of the Apollo mission had many derivative benefits to society, benefits that are not predictable. Certainly the advances in electronics, the development of integrated circuits, and the changes that ensued are attributable to advances created by government funding of the Apollo mission. This includes personal computers, high-definition television, and communications. Jerry likes to point to the contract Corning was awarded for $5 million to develop ceramic heat shields for the rocket on reentry into the earth's atmosphere. This required research and development for a material that could withstand the thermal gradients and mechanical vibration accompanying reentry. A commercial derivative of this technology is Corning-ware, ceramic pots that you can use for cooking. Thus, the government's investment in the heat shield paid off in several ways, not only in protecting the rocket and its personnel, but also in creating a stronger commercial company.

Engineering is very satisfying career, Jerry notes, where you participate and improve the quality of life of people by solving problems with solutions that work well. In the United States there has always been a search for the new frontier, typified by the Western migration. Whereas there is no literal frontier these days, engineering may be the last frontier, a virtual frontier. The Internet, the frontier of the moment, and its evolution, the frontier of the future, are and will be frontiers created by engineering invention and innovation.

Jerry recommends initially working for a large company for three to five years to gain experience and learn in what is very often a teaching and learning mode. You will meet a variety of people, developing practical applications to your engineering knowledge, solving problems including how to manufacture your solutions. Universities provide the knowledge base for problem solving; but real-world engineering cannot be learned in a classroom, it must be experienced. Budget and time constraints are real, not theoretical considerations, and their effects direct the designs you create.

Teamwork is very important. For instance, the initial design of a printer circuit board had a very poor mean time between failures (MTBF) of 50 h. A team analyzed the board in light of manufacturing and quality control techniques, recommending design modifications that increased the MTBF to 5000 h. The changes were not made to the electronic conceptualization of the design, but rather in ensuring that manufacturing and quality control checks could be completed as accurately as possible. In the initial design, components faced in a variety of directions; it was something the design engineer did not consider important in specifying. However, for quality control checking, it is much easier to find a component misattached if the directions and hence polarities are the same. Similarly in manufacturing, care was taken so the machine providing the autoinsertion of parts did not have unnecessary and time-consuming rotations and gyrations. Production capacity increased dramatically from just this one consideration. The layout now allowed ready quality control checks, and the resultant printer garnered much of the market. Teamwork, as manifested in concurrent engineering, not only allows all the groups to work together on initial and subsequent designs, but also facilitates quantum leaps in design improvement rather than the incremental improvements occurring from one year to the next in the traditional design process.

Expand your view of what communications means. Jerry points out that communication skills are important, particularly sales skills. You must sell yourself and your ideas to management, competing for resources to develop your ideas, not those of another person. Personal appearance matters, too; it is part of what you are communicating. Writing is an essential part of engineering; you need to write progress reports and memos. A very useful communications concept to remember as you draft these reports and memos is what Jerry refers to as the *doctrine of complete action*. The report or communication should contain all the material necessary for someone not directly involved with the project to make a decision about the project. In the case of a memo, if you refer to a customer problem, attach the customer's letter to your response so that your superior knows what the problem was and how it was resolved. In the case of a report, you should describe the problem and solution as if the person reading the report were not aware of the problem specifics. Your immediate superior will be aware of the specifics, but the report may be passed on to others who lack this knowledge. There should be sufficient information for those people to also make an informed decision. In report writing, summarize the problem and solution in the beginning; as the report moves to higher echelons, it will probably not be read in its entirety, but the summary information will be.

Quality Assurance

Quality assurance procedures are created to make sure the product is produced according to plans and meets the specifications called for, such as size, resistance, and weight. The quality standard

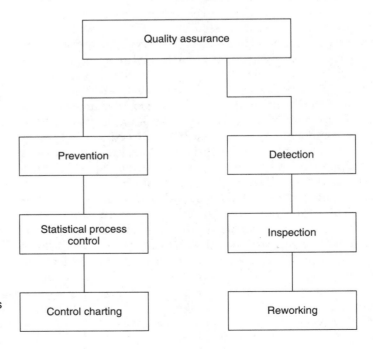

Figure 5.2
The two basic components of quality assurance are prevention and detection.

indicates the variation allowed in the final product. A tolerance analysis anticipates what the variations may be; but in the manufacturing process there may be a deterioration in machine performance, or the workers may vary in the precision with which they fabricate a part. Controlling this variation is the goal of quality assurance.

This is accomplished with prevention and detection, as indicated in Figure 5.2, at the bottom of page 168. Prevention means to prevent variation in materials or processes before parts are made. Detection means inspection of finished parts after they are produced; of course, it is wiser, more cost-effective, to prevent rather than to detect errors. As workers perform a task, there will be a variation in the output characteristics of the product; these are accounted for in the tolerances set in the design process. Statistical quality control is one way workers take control of the production process to ensure the quality of the output. Assume that the impedance of a circuit should be 100 Ω with a tolerance of 1 Ω. Production line employees will sample every 100th circuit for its impedance, plotted in Figure 5.3, keeping an eye on the performance. As long as circuits produced by the machine fall within the tolerances specified, no action is required. Should the circuits fall outside the tolerance band, production is stopped and adjustments are made to the machine.

Once the product is manufactured, it is tested to see whether it meets the customer specifications—materials, size, function, and performance. When there are very large production runs of devices, it is prohibitive to test every unit, so statistical sampling is done here as well. For smaller runs and those with more complex devices, each unit may be tested. For instance, with electronic devices, there is a *burn-in* time. Electronic devices that fail tend to do so in the first few hours of operation, so manufacturers will run the devices for a few hours, and if the

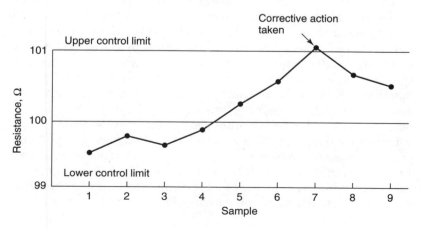

Figure 5.3
A control chart is used to track the manufacturing process.

device meets the specifications after the trial, it will probably last a long time.

Long Island Railroad's Doug Haluza

Doug Haluza is an electrical engineer with 15 years' experience in the communications field. He moved into engineering management a few years ago while retaining some design engineering responsibilities, and thus he provides a perspective from both vantage points. He always wanted to be an engineer, tinkering with batteries, lights, and switches since kindergarten. At one point, he considered being an attorney, but found that in law, winning matters, not truth; in engineering, truth matters. It is important that a design be safe and that it function correctly.

His first job upon graduation was with a firm which won a contract to build a low-level wind shear alert (LLWSA) system for the Federal Aviation Administration. It was a fairly simple design in which five anemometers, located at selected points around an airfield, measured the local wind velocity. When there was a vector difference of 15 knots (kn) between two of the anemometers, a wind shear was assumed to exist, which is a potentially dangerous situation for aircraft landing and taking off, and an alarm sounded. The original design was produced under government contract with a different company, which of course anticipated winning the second contract, but it was not the low bidder. Doug's company used newer technology in terms of personal computers to link the anemometers and to perform the wind shear calculations; the company also redesigned the subsystems. Thus, the company reduced the number of chips by one-half compared to the original design, reduced the computational cost, and was able to absorb the engineering design costs and still make a profit while winning the low bid. The company that originated the design did not continue to innovate, to improve the design in light of technological advances, staying with a comparatively expensive design.

This scenario is one that Doug has seen played out several times: the need to maintain technological currency in times of high technological change. The pressure on managers to perform in this environment is also great, as you want to pick correct future technologies, developing a technological intuition. He illustrates this with a situation we are more familiar with. Several years ago there was a choice in videocassette recorders (VCRs) between Betamax and VHS standards for video recording. VHS technology became the standard; companies that invested in Betamax had to retrace their steps and reinvest in VHS. Managers worry about capital (money), human resources, and time. They sometimes find that even if the capital can be recovered from investing in an incorrect technology, the time lost to competitors may be the more costly error. In many organizations this is further complicated

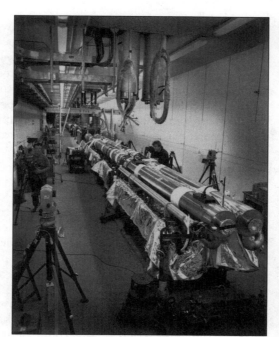

Photo 9
Superconducting magnets are essential to hold the highly accelerated ions in the Relativistic Heavy Ion Collider. *(Courtesy of Brookhaven National Laboratory)*

by the elimination of many middle-management positions which provided technological control to more senior management; the latter group now must have the ability to change and adapt.

Traditionally there has been a reluctance to change on the part of people, be they engineers or managers or both. One characteristic of the information age we are entering is the recognition that major changes in the knowledge base will occur, not once, but several times. Such a change occurred in the 1960s when vacuum tubes were replaced by transistors. Engineers who knew a great deal about vacuum tubes, the subtleties of designing with them, found these skills wanting as transistors proved to be more reliable and less costly. The engineers needed to continue their education, be it with new degrees or professional education courses so they could remain effective design engineers. The trend continues in all fields of engineering; what you learn in college and apply in the first few years in engineering will not be sufficient for your career. Additional education will be the norm.

Doug also learned the importance of the design notebook or journal on his first job, when he started to design a power supply by physically connecting components in the laboratory and seeing whether they met the specifications. Swift and not too subtle direction from the president and chief engineer pointed out the need for calculations in a bound design notebook, where additional thoughts and comments were also kept. The company wanted to show that current design standards were followed, should problems develop in the future, or, more positively, that the notebooks provide documentation for a patent application.

When questioned about attributes he would seek in an entry-level engineer, Doug wanted more than technical competence—that plus indications that book theory could be blended with hands-on acumen. He certainly wants a person who can be promoted, who demonstrates initiative. Communicative ability is very important. An engineer needs to stand up and be heard, to convince others to get resources vital to a project. To fail to articulate a potential problem and have it develop is not good for a company, and not wise for career progress. Anticipating problems is an important part of an engineer's job; and while it is impossible to predict when or where a derailment will occur because someone backs up over a switch, for example, designs must anticipate this problem. Written communication is vital as often there is not time for face-to-face discussions with a supervisor, and reports and memos play a major role in the communicative structure.

The railroad views itself as having many systems, subsystems, and components. So an engineer who looks at only components, missing the interaction of the component with the rest of the system, is not performing the job correctly. This can be as seemingly mundane as replacing speakers in a public address system, where a new speaker is incompatible with the others and upsets the system balance. While messages from it can be heard very well, other speakers become inaudible.

Safety and maintainability are two important features for an operating company, where equipment needs to be repaired and designs should allow for ready and safe repair. Automobiles have been designed in which the engine had to be pulled to replace the rear spark plugs; insufficient attention was paid to maintenance, increasing the life-cycle cost.

The expense due to accidents is huge—costs for personal injury, repair, and loss of equipment. Safety needs to be designed into devices, not as an afterthought that can be added like a bandage. Exposed wiring needs to be insulated and encased so that a casual nick during an overhaul cannot cause a failure. Asking yourself whether it can be done differently or more safely is an important question, one you could be asked to defend in a lawsuit if an accident occurs.

This returns us to accepted practice, design to accepted standards not your own, unless your standards include and surpass the accepted standards. You can use the experience of others in extending the designs through creativity. One of the most successful technological efforts in modern times is the Apollo mission to moon in the 1960s. While many scientific breakthroughs were achieved, there were many more incremental design improvements, resulting from learning from previous designs and then innovatively extending this knowledge. The rocket itself followed this path of design innovation.

Great Barrington's Bruce Collingwood

Bruce Collingwood is the town engineer for Great Barrington, Massachusetts, arriving at his current position after nearly a decade of working for civil engineering consulting firms. Before discussing his current position, he reflected on the somewhat circuitous path that brought him here. As a high school student, he was interested in chemistry and considered pursuing a career in medicine, but that goal faded after his first year in college; and he thought chemical engineering would be the most intriguing, challenging, and personally rewarding path to follow. By the end of his senior year, he realized that working for large companies, the most probable career path in chemical engineering, did not interest him as much as civil engineering, where he could work in smaller consulting firms. Bruce decided to pursue a master's degree in civil engineering, filling in gaps in his undergraduate education as he did so. Following graduation he worked for a consulting firm specializing in geotechnical engineering, confronting problems as varied as golf course hydrology studies, designing earth structures, and performing foundation studies.

Before a commercial building is constructed, the soil is tested to ensure that its load-bearing properties will safely support the proposed building weight. Soil samples are taken and analyzed, and calculations performed to determine the type of building foundation for the site. Engineering analysis and a report to the client are part of the process. One of the challenges Bruce enjoyed was the work variety that consulting provides. Seldom are two jobs the same, so there is always the opportunity to creatively apply your knowledge to new situations. There is also a role for engineers who like to specialize in a certain topic, perhaps water treatment, and become the resident expert. These engineers work on other projects, to be sure, but will always be called upon for their forte.

As with most professions, engineering has cyclical employment opportunities as companies depend on product sales or contracts to fund the workforce. Consulting engineering firms are no exception, and Bruce found himself in the unenviable position of being in middle engineering management—not quite in senior management, but several years beyond the entry-level area. It is this middle area that tends to be cut first in downsizing, and Bruce did not escape the final round as fewer and fewer contracts were secured in a time of economic contraction. However, being observant and noticing the problem of fewer contracts, he had begun to search for another position; so when he was terminated, he transitioned to a branch office of another civil engineering consulting firm near his hometown. He recommends that you continually keep an eye on the engineering marketplace, be knowledgeable about the business, and maintain your professional connections.

There were only two engineers in the branch office, so he had expanded opportunities to not only perform work but also attract new clients. How does a small firm obtain business? Where do the contracts come from? The answers lie in the many forms of community involvement. You may become active in civic organizations, such as Kiwanis or Rotary clubs, or join and participate in organizations whose members use consulting services, such as town managers' associations. You establish a rapport with people over time, perhaps attending community events together, and from this rapport recommendations will come as the need arises. Another avenue, or simultaneous avenue, to pursue is to volunteer to serve as an adviser to local governments or civic groups.

Because his company was the consulting engineering firm on a large waterworks project, when the owners of a small waterworks asked local officials whom they would recommend, Bruce's name and company were at the forefront. Bruce drew up a proposal for them, outlining the various design steps that needed to occur as well as financial requirements and projections. The waterworks agreed, and a contract was signed based on the proposal items.

The cooperative waterworks system was really quite simple: a pump, drawing water from an artesian well, discharges into a pressurized tank. Distribution piping leads from the tank to the various homes served by the waterworks. The problem was that the tank was failing, leaking badly and needing replacement. The Massachusetts Department of Environmental Protection (DEP) requires that, when water systems are replaced, they be upgraded to comply with existing regulations, thus requiring an engineer to design the replacement tank system. Bruce had never designed such a system before, but his education provided him with the methodology necessary to undertake the design. First, he acquainted himself with the site and performed a reconnaissance of the waterworks, noting the various components of the system and the houses served by it. This provides the data necessary for subsequent calculations. Many of the calculations and components are determined by DEP regulations for water supplies. Figure 5.4a shows a schematic of the system, and Figure 5.4b is a sketch of the tanks and related equipment, which now includes air compressors to maintain tank pressure and filters to remove sediment from the water. The previous system had neither. The engineer's job does not end with the plans. The DEP has to approve the plans, and once they are approved, a contractor is sought to install the equipment. The engineer provides the technical requirements for the contract, which is finalized by an attorney. Maintaining contact with the DEP engineer is important because it facilitates the approval process, resolving questions quickly. The engineer provides some oversight to the contractor's installation, making sure

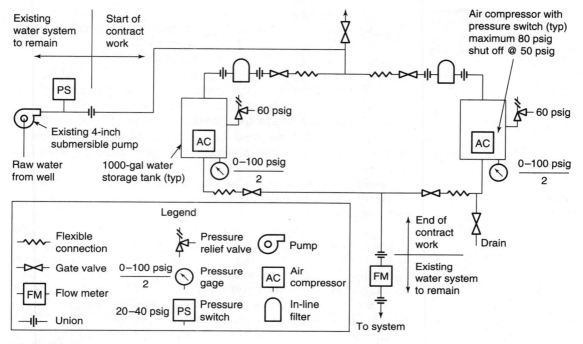

Figure 5.4a
Water piping schematic.

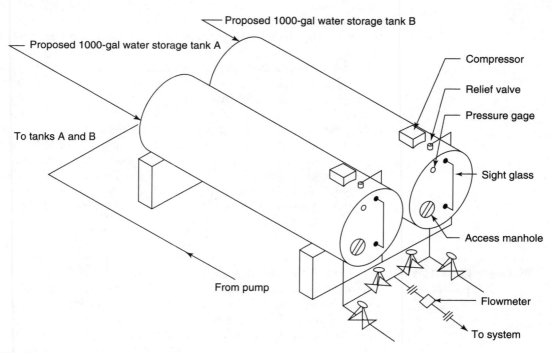

Figure 5.4b
Water system.

that the approved design is the one installed and not a variation of it. You can deduce that most successful consulting engineers have good interpersonal skills, necessary for working with clients, government agencies, and contractors to satisfactorily complete a project. Of course, on larger projects and in larger organizations, you will not perform all the tasks. That is, indeed, one of the trade-offs in working for larger or smaller firms.

Always keeping an eye open for opportunities, Bruce found that Great Barrington, located adjacent to his hometown, was looking for a town engineer with a background that fit nicely with his. Another career path now lies ahead of him. In this position, his engineering design and analysis knowledge is not as important as his ability to coordinate the work of two dozen people, repairing and restoring highways and wastewater systems and performing building maintenance. However, his problem-solving skills, honed by years of design and analysis experience, are called upon every day. He works with contractors and may be in a position to recommend the hiring of consulting engineering firms for various projects, such as building renovation. He will be networking with other town engineers and public works superintendents, sharing knowledge and building support systems.

An engineering journal carries forward in Bruce's life; even though he does not deal with design work, he now maintains two bound journals. One is a notebook that he carries with him, noting repair activities, concerns and wishes of town officials and residents, and reminders for new projects. At the end of day, he transfers these notes into the second journal, where there are different categories that he adds to and annotates daily. He marvels at how new technology appears in the most unusual places, indicating once again the need to keep technologically current through continuing education. Surveying has been tremendously helped by the global positioning system (GPS), in which the coordinates of a point on the earth's surface can be determined via satellite, rather than through line-of-sight use of the transit. Backhoe operators use a laser level to determine how deeply to dig a trench, attaching the receiver to the backhoe and the laser on the ground at the appropriate level. When the backhoe reaches the appropriate depth, the receiver emits a sound, indicating to the operator to cease excavation.

Sidney Bowne's Carl Becker

Sometimes the initial attempt at engineering is not successful, but maturity and changing circumstances permit success. Carl Becker, an engineer in a medium-size civil engineering consulting firm, Sidney B. Bowne and Son, specializes in water supply systems, their design, maintenance, and environmental compliance. When he graduated from high school with excellent mathematics and science abilities, he thought a career designing airplanes and rockets would be exciting and rewarding. However, the one-to two-hour commute each way from home to school plus studying with many distractions from younger siblings and family members proved to be disastrous, and he quit before being terminated from college. Carl was able to secure a job as a drafter because of technical drawing courses he had taken in high school, and he reenrolled in the engineering school he had unceremoniously left a few years earlier, restarting his engineering career on a part-time basis. This time he excelled; he was able to study in a quiet environment, and the commute was much less, as he had his own apartment. His view of engineering also had changed, and he was interested in environmental engineering, known as *sanitary engineering* in those days.

Upon graduation from college, he went to work for a small engineering consulting firm where he was involved in a wide variety of projects, a course of action he highly recommends for today's beginning engineers. A wide and varied experience base helps you decide where you might like to specialize, say, in structural design, water supplies, or highway design. After several years of experience, Carl found that water supply design was the area he wanted to focus on, and he has maintained this interest, with mounting expertise, for over two decades. His education did not stop; he achieved a master's degree in water quality engineering and studied for and passed the examination for a professional engineer's license. This license is essential for the private practice of engineering, such as in consulting firms, and is highly recommended for those working in large corporations.

He moved into middle management of the firm, which had expanded from 10 to 30 engineers with the resultant need for managerial guidance as well as mentoring of the newer engineers. The expansion also increased tensions between the old guard and new; and Carl, sensing intrigues, moved to a local water supply company as its assistant vice president of engineering. He quickly became chief engineer and then vice president of engineering and actively moved to upgrade a system that had fallen into disrepair because of more than a decade of austerity in the company. He and the company's president shared a similar vision for the company: to turn an old, manually operated facility into a modern, automated one. There were personnel, financial, and technical problems to solve; it was a challenging and rewarding time. The president was promoted to a higher position within the parent corporation and no longer had an association with the water company. In his place a finance executive from the parent company, with little water company experience, was installed. The new president's views were not consistent with those of professional engineers. When one of Carl's annual reports to the Public Service Commission was altered by others, Carl refused to sign it and knew it was time to move on. His stature in water supply and water quality was widely recognized and enhanced through his participation in professional organizations, serving on technical committees, and being an officer in the regional chapter. A lunchtime meeting he thought was going to be a pitch for a new service turned out to be an offer to join Sidney B. Bowne. After amiable and quick negotiations, he signed on as an associate partner and has found a place with values he respects and challenges he desires.

Carl finds that the engineering workday often falls into thirds— one-third engineering design and analysis, one-third accounting and budgeting, and one-third communications, oral and written reports, letters, and memos. Reflecting on his undergraduate education, he believes the liberal arts component is very important, particularly as it develops communicative ability and the ability to work with others. Enhancing your interpersonal abilities, working with others, will assist you on team projects and in meetings with clients. Much engineering design and analysis is accomplished with computer software, so being proficient in computer-aided design is certainly important. Your technical expertise provides essential bottom-line credibility.

Another quality he values, although it is difficult to measure, is common sense. After a project has been carefully designed and the construction phase begins, unanticipated situations will arise which require design adaptations. Engineering intuition, common sense in this context, will give you a sense of when something can or cannot be done. There is pressure to allow design alterations to move the project along; personnel and equipment sit idle while costing money until the redesign is approved. Negotiations with contractors as they seek changes are important in facilitating a successful project. Knowing when a redesign is required, when a field change can be safely made, is a skill that can be honed, but it rests on the bedrock of common sense. In college this intuition can be developed by paying attention to problem answers. Do they make sense, based on the inputs? Do they seem correct? This type of questioning strengthens your intuition.

Carl illustrated how common sense plays a role in engineering designs. A local water supply company had to replace an elevated water storage tank and to expand system capacity. It contracted with a competing engineering firm to design the new tank, which it did according to codes. The solution involved maintaining the existing elevated tank in service to provide the required system pressure while a new elevated tank was constructed above an existing ground storage tank on the same site. This solution precluded a method to dismantle the existing tank and required a costly reinforced-concrete tank to be constructed to carry the weight of the elevated tank above. The plans were completed, and the project was bid to contractors. The response to the bid placed the project's cost at twice the budgeted amount. The water supply company turned to Sidney Bowne and hence to Carl, who suggested and then designed a system that would maintain system pressures without the elevated tank. This allows the existing elevated tank to be dismantled and a new elevated tank to be constructed on the same location. In this situation, a ground storage tank with variable-speed pumps was constructed to maintain the water system pressure while the elevated tank was out of service. The ground storage tank was needed to eliminate a deficit in the water supply company's storage capacity. The new design had a construction cost about one-half that of the initial one and fit within the company's budget.

Engineering is very rewarding—it is great to solve problems, to see your designs built, Carl points out. It is also demanding and requires constant skill and knowledge development. Sometimes a project will come along that requires you to relearn material, typically an option exercised at night, at home. Equipment suppliers also provide workshops; in the water supply business a new product may allow a different system design. The workshops are technical ones, and while they introduce the product, they are designed for engineers who may want to use the product and have to be able to design systems incorporating it. Professional journals indicate trends, as do technical presentations at professional society meetings. You need not be on the cutting edge, but neither do you want to be at the back of the pack, among the last to learn and incorporate new technologies. Again, your technical expertise is the foundation of your profession.

Long Island Lighting Company's Donna Tumminello

Donna Tumminello works in the research and development division of the Long Island Lighting Company (LILCO), where she conceptualizes the need for and then commercializes new products. It is a very exciting position and seemingly unusual for a public utility, yet it makes perfect sense as, Donna explains, different LILCO departments face technical challenges that need new technologies to overcome them. Research and development (R&D) initiatives are directed toward solving these problems, through contracts with Long Island research and business communities. The outcomes of these contracts sometimes result in new patents and products of which LILCO shares part of the future profits. Donna thrives in the competitive atmosphere that seeks to carry the technology used to solve a LILCO problem to its full potential, resulting in a new product or service.

Donna started with another company in the mid-1980s, working part-time throughout her undergraduate days. The idea of a co-op experience, or technical work experience as an undergraduate, provides a very worthwhile lens for viewing the profession. It also lent credibility to the courses she took and provided insights into some of the physical manifestations normally appearing as equations in books. She worked at the branch office of a company involved with combustion and combustion control, her specialty being computerized combustion control. This company solved combustion control problems for others, often responding to requests for proposals (RFPs) and, when successful, completing the work as detailed in the proposal. Sometimes this required all-nighters to get the bid out on time, creating close personal friendships that sustained the, at times, high-pressure work. The office manager charted the dollar value of contracts, so all could share the excitement of increasing clients and the positive slope to the chart line and the distress if a change of direction in the line occurred. The bottom-line focus of the engineering business persuaded Donna to obtain a master's degree in business administration, specializing in finance. The early 1990s was a time of mergers and acquisitions, and her company was acquired by another. Of course, this raised concerns as to whether the employees would be moved, merged, or relocated as a result. Noting that LILCO was searching for an engineer to create and oversee RFPs such as she had been responding to, Donna applied and won the position. She very much likes the comparative financial stability that a public utility provides, and her entrepreneurial streak is satisfied by facilitating others to create new products.

Donna and her identical twin sister grew up in a technical household; her father was an engineer, her mother a nurse, although engineering was not pushed as a career choice. In fact, her father argued against it, as he thought it was too male-dominated, it would be difficult to raise a family, and there was little potential to work part-time in engineering. Donna was proficient in physics and mathematics and had a practical mindset, while her sister preferred the life sciences and chemistry and attained a pharmacy degree. She credits a creative high school physics teacher for convincing her that engineering was the best career choice for her. Donna attended a nearby community college before completing her last two years in Hofstra's electrical engineering program. While she was in the minority in most engineering classes, she never felt isolated because she was a woman and finds that she continues to share the interests of most of her male engineering colleagues. She has never felt that gender was an issue in her career, although she agreed that in the last months of pregnancy she did feel a little differently, working until the day before her daughter was due. The question of child care now enters the picture, and good fortune assists as her mother-in-law decided to become a child care provider a few years ago and now will perform that function for her granddaughter, resolving a potentially troubling decision for Donna and her husband. It remains very difficult for many to combine child care and full-time work as a professional in engineering, law, medicine, or business. Part-time employment is often not available and, if available, may not provide the desired career path. Donna has always balanced school and work and views work and family as a new challenge, but one that she and her husband can satisfactorily meet.

Donna likes to be busy, and even though her first job was very challenging and time-consuming, the company offered full tuition compensation for advanced degrees. The idea of getting a master's degree was not something she planned on, but free tuition was too good a deal to pass up. Looking back, Donna thinks an undergraduate engineering degree plus an MBA is a dynamite combination. An engineering program does not give you a business sense, the bottom-line attitude that is essential for a successful company. For instance, she always performs a cost/benefit analysis before committing R&D funds only to projects that are positive. An interesting idea is not sufficient; it must also have a good benefit/cost ratio.

Most of the advanced-level courses in her MBA program were team projects involving investigation, working together on the conceptualization and analysis phases and a joint presentation to the class where incisive questions were the norm. The analytical aspects of engineering provide an excellent base for the degree, but other attributes are needed as well. The ability to work well with people of diverse aptitudes is important, and respecting those aptitudes is equally vital. She finds this particularly the case in R&D, where the projects are wide-ranging. For instance, in one class, people were asked to describe time. Donna's response, typical of that of many of her engineering colleagues, was a clock, morning, noon. A colleague from marketing indicated time as being of the essence, timeliness. A good manager respects and utilizes these different attitudes and abilities in a creative team.

Her finance courses were challenging and very interesting, situations in which her competitive sink-or-swim attitude always had her swimming quickly; her marketing courses presented a different challenge, and she adopted a swim-with-a-buddy system to prevent sinking. She finds that marketing people are witty, creative, a little "off the wall," and not at all like most engineers, and the courses reflect this personality. One of the challenges her marketing team faced was to develop a new marketing strategy for a major company undergoing difficult times in the marketplace; their choice was Ben & Jerry's ice cream company. They developed a plan that won kudos in class and from the professor. With her "take it to the limit" attitude, Donna persuaded her team to present the plan to Ben & Jerry's chief executive officer (CEO). Several members of the team agreed, and they wrote to the CEO, who was impressed and invited them to make a presentation—an exciting culmination to the course. This attitude of taking it to the limit, don't stop with your notebook, is one that she finds essential for success in today's business world.

Her experiences in and out of school have convinced her that linking marketing types and creative technical engineers with good ideas is the most likely combination for creating a new start-up company. It is too much for one person to handle the technical nuts-and-bolts issues of engineering design and manufacturing and to also conceptualize, plan, and execute sales and marketing strategies. Not surprisingly, Donna has thought about business ventures of her own, but right now with a newly expanded family, such thoughts are in the background.

Donna finds that engineering is making a comeback, and many job opportunities are opening up for engineers with a bottom-line attitude. Her experience indicates the need to keep an eye on the ball, as mergers and changes in market conditions affect company staffing levels. There is not the view of lifetime employment any longer, hard work matters, and expectations are greater for engineers in the workplace. She makes the comparison to figure skating. Fifteen years ago, a double spin was considered top form; today a triple spin is required for anyone who expects to compete near the top. An additional degree may be viewed as the required extra spin for those who want to move up the career ladder.

References

1. Arkebauer, James B. *The McGraw-Hill Guide to Writing a High-Impact Business Plan: A Proven Blueprint for First-Time Entrepreneurs.* New York: McGraw-Hill, 1994.

2. Lechter, M. A.; E. C. Clifford; and R. B. Famiglio. *Successful Patents and Patenting for Engineers and Scientists.* New York: IEEE, 1995.

3. Siegel, E. S.; B. R. Ford; and J. M. Bornstein. *The Ernst and Young Business Plan Guide,* 2nd ed. New York: Wiley, 1993.

Appendix A
Problem Solving

One of the wonderful attributes of an engineering education is the ability to solve problems; however, one of the difficulties that many engineering students have is the solution of algebraic word problems. Much of engineering analysis involves the solution of just such problems, hence algebraic facility is necessary for the successful study and practice of engineering.

The problems in this appendix are not meant to replace a course in precalculus or algebra but will provide a quick review, with problems drawn from fields of engineering. The engineering analysis methodology is used, consistent with Chapter 4 and most engineering texts.

Linear Equations

A linear equation is one of the form $ax + b = 0$, where a and b are constants $(a \neq 0)$ and x is the variable. The solution of the equation is $x = -b/a$, quite easy. The challenge lies in translating the word problem to a linear equation, which is not so easy.

Example A.1 In a chemical processing plant, a mixture containing 30 percent alcohol is obtained by adding a 60 percent alcohol solution to 40 kg of a 20 percent alcohol solution. Find the kilograms of the 60 percent solution required. All percentages are on a mass basis.

Solution

Given:

A known mass of an alcohol solution, the percentage of alcohol required, and the percent alcohol of mixture to be added.

Find:

The amount of a 60 percent alcohol solution that must be added to obtain the correct final solution.

Sketch and Data:

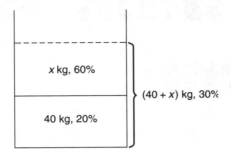

x kg, 60%

(40 + *x*) kg, 30%

40 kg, 20%

Figure A.1

Assumptions:
Conservation of mass of alcohol.

Analysis:
Let *x* be the amount of the 60 percent solution added. The total number of kilograms is 40 + *x*. The mass of alcohol in the mixture is the percentage times the total mass. From this assumption we postulate that none of the alcohol can disappear. Thus, the conservation of alcohol mass is

Initial alcohol	+	Alcohol added	=	Final alcohol
$(0.2)(40)$	+	$(0.6)(x)$	=	$(0.3)(40 + x)$
8	+	$0.6x$	=	$12 + 0.3x$
		x	=	13.33 kg

Example A.2 Three pipes may be used to fill a tank. A student measures the times to fill the tank for each of the pipes and records the following data: 20, 15, and 40 min. Determine the time to fill the tank using all three pipes simultaneously.

Solution

Given:
A tank and the times to fill the tank individually from three different pipes.

Find:
The time to fill the tank using all three pipes simultaneously.

Sketch and Data:

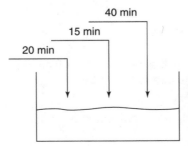

40 min

15 min

20 min

Figure A.2

Assumptions:

The flow through any pipe does not change when all are used simultaneously.

Analysis:

Let the time to fill the tank be t min. In 1 min the amount of liquid entering the tank is

$$(1) \left(\frac{1}{20} + \frac{1}{15} + \frac{1}{40} \right)$$

Note that the first pipe fills 1/20 of the tank in 1 min, the second pipe fills 1/15 of the tank in 1 min, and the third pipe fills 1/40 of the tank in 1 min. If we add the percentage expressed as a decimal, that each contributes per minute, $0.05 + 0.0667 + 0.025 = 0.1417$, we find that 14.17 percent of the tank is filled. Thus in t min, the tank will be totally or 100 percent filled.

$$(t)(0.1417) = 1.0$$

$$t = 7.06 \text{ min}$$

Problems

A.1. A student received grades of 85, 84, and 91 on her first three chemistry quizzes. What grade must she obtain on her fourth quiz to reach a 90 average?

A.2. A student received grades of 78, 82, and 72 on three tests in physics. The final examination counts as two test grades. What must she score on the final examination to have an average of 80 for the course?

A.3. The octane rating of a gasoline is determined by comparing an engine's peak pressure from an actual gasoline mixture to a standard value. The octane rating of a mixture is determined by the volumetric addition of the fuels; thus equal volumes of 80 octane and 100 octane yield a 90 octane mixture. Determine how many gallons of 95 octane fuel must be added to 100 gal of 85 octane fuel to obtain a 91 octane mixture.

A.4. The pumps at a service station blend 87 octane with 95 octane gasoline to obtain an octane rating between the two. A customer receives 15 gal of 92 octane fuel. How many gallons of 87 octane were used?

A.5. A tank holds 500 kg of brine with a salt concentration of 20 percent by mass. How much water must be evaporated so the concentration rises to 50 percent?

A.6. The radiator in an automobile holds 4 gal of a 10 percent antifreeze/ water mixture. The percentage of antifreeze must be raised to 25 percent by draining some of the mixture and adding 100 percent antifreeze. How much mixture must be drained? All percentages are on a volume basis.

A.7. A ceramic clay contains 50 percent silica, 10 percent water, and 40 percent other minerals. Determine the percentage of silica on a dry (water-free) basis.

A.8. Gold has a value of $12 per gram (g). A student finds a large gold ore nugget weighing 1000 g that contains gold and quartz. The density of gold is 19.3 g/cm³, the density of quartz is 2.5 g/cm³, and the density of nugget is 6.5 g/cm³. The student is offered $150 for the nugget; should he accept the offer?

A.9. A tank may be filled using pipe *A* or *B* with times of 10 and 20 min, respectively. It takes only 5 min to fill the tank when pipe *C* is used simultaneously with pipes *A* and *B*. How long does it take to fill the tank using only pipe *C*?

A.10. Two workers, *A* and *B*, can assemble a device in 3 and 5 h, respectively. How long would it take to assemble the device if they worked together?

A.11. An airplane flies with a velocity of 250 mi/h when there is no wind. In flying with the wind, it travels a certain distance in 4 h. However, in flying against the wind, it can travel only 60 percent of that distance. What is the wind's velocity?

A.12. A state's automobile inspection program checks headlight alignment with the specification that the light beam drop not be greater than 2 in. for each 25 ft in front of the car. Suppose that the headlights on your car are 28 in. above the ground and that they meet the 2-in. drop requirement. What is the minimum distance in front of the car that they can illuminate? If you are driving at 50 mi/h, how long does it take to travel that distance?

A.13. A tractor has a belt pulley diameter of 10 in. operating at 1100 revolutions per minute (rpm). The pulley is connected to another machine that needs to operate at 650 rpm. What size pulley should be used on the machine?

A.14. The following table lists the amounts of grain and hay that a steer is fed to produce the desired weight gain. The grain costs $0.15 per pound, and the hay costs $0.06 per pound. Determine the lowest-cost combination.

Combinations of grain and hay to produce
satisfactory weight gain in a steer

Pounds of hay	Pounds of grain
1000	1316
1100	1259
1200	1208
1300	1162
1400	1120
1500	1081
1600	1046
1700	1014
1800	984
1900	957

A.15. Inventory turnover is the ratio of the cost of goods sold to the average inventory value:

$$\text{Inventory turnover} = \frac{\text{Cost of goods sold}}{\text{Average inventory value}}$$

XYZ Office Supply Company started the year with an inventory worth $28,532 and ended the year with an inventory worth $33,124. During the year the business sold $264,845 worth of office supplies. What is the inventory turnover ratio? Is a high ratio desired? Why or why not?

A.16. The standard aspect ratio (width to height) for a television picture is 4:3. The program director has scheduled to broadcast a wide-screen movie that was filmed with an aspect ratio of 2:1. Suppose that the showing will broadcast the full width of the movie, while maintaining the aspect ratio of 2:1. What would the result look like on the television screen? Illustrate with a sketch. What would happen if the movie were broadcast at full height?

A.17. You are managing a mail-order business that has five workers assigned to process and package orders. These workers are packaging an order for 2500 parts which needs to be mailed by the end of the 8-h shift. So far they have packaged 900 parts during the first 4 h. It is evident that they are not going to be able to finish the order during the time remaining without additional help. How many additional workers are required to complete the task during the last 4 h?

A.18. During the first year of operation, a clinic treated a total of 4916 patients and gave 624 of them influenza immunizations. This year the clinic is treating more patients, 3384, during the first 6 months and has given 487 flu immunizations. You need to order supplies; estimate the number of patients and flu immunizations for the rest of the year.

A.19. A water treatment plant deals with very large volumes of water and very small concentrations of chemicals used in the water purification processes. In one process a chemical must be added in the amount of 0.7 parts per million (ppm). The treatment facility has holding tanks that contain 850,000 gal. How many gallons of chemical should be added to the holding tank to obtain the desired concentration?

A.20. A 2-in.-diameter pulley is mounted on a motor shaft that rotates at 1200 rpm. This pulley is connected to a 5-in.-diameter pulley mounted on the shaft of a fan. Determine the fan's rotational speed.

A.21. The gears in a transmission are often described in ratios, such as 3:1, which means that the drive gear rotates 3 revolutions (rev) while the driven gear rotates 1 rev. In a certain automobile, the final transmission and differential gear reduction is 3.2:1. The engine crankshaft rotates at 2600 rpm. How fast does the drive axle rotate? If tires with a 26-in. diameter are attached to the axles, what is the speed of the car at this rotational speed?

A.22. A worm gear is often used in manual winches. The worm gear has a ratio of 45:1, indicating that 45 turns of the hand crank are needed to rotate the large gear once. Suppose that the drum attached to the large gear has a diameter of 2.5 in. and is used to reel in 12 ft of cable. In a test, you find you can rotate the hand crank at 40 turns per minute. How many minutes will it take to reel in the cable?

A.23. The pressure in a system may be increased by using two pistons of different diameters. For the pistons to remain in static equilibrium, the forces on each piston surface (pressure times area) must be equal. The larger piston has a diameter of 10 cm, the smaller piston has a diameter of 6 cm, and the pressure acting on the larger piston is 140 kN/m². What is the pressure acting on the smaller-diameter piston? If the diameter of the larger piston increases by 20 percent, everything else remaining constant, what is the pressure on the smaller piston?

A.24. The steepness of a railroad track over a 3-mi grade is reported as a rising grade of 1 in 43, meaning that it rises 1 ft for every 42 ft in track. Following the rise, the track now descends 5 mi with a descending grade of 1 in 79. Is the elevation at the end of the 5-mi descent less than, greater than, or equal to the elevation at the start of 3-mi ascent?

A.25. The compression ratio of an internal combustion engine is described in terms of volume ratios, the volume at the beginning of the compression divided by the volume remaining at the end of compression. These compression ratios are expressed as 8:1 or 15:1. A diesel engine has a cylinder of with a 6.25-in. bore and an 8-in. stroke. The volume left at the top of compression stroke is 15.9 in³. Determine the engine's compression ratio.

Simultaneous Linear Equations

Consider the equation $ax + by = c$, where x and y are variables and $a, b,$ and c are constants. This is a linear equation in two unknowns. If there are two equations such as

$$a_1x + b_1y = c_1$$

$$a_2x + b_2y = c_2$$

and they are independent of each other and consistent, they may be solved simultaneously to determine the common value of x and y. This may be done by addition and subtraction, substitution, or use of graphics. When the equations are plotted, their intersection provides the solution. Dependent equations form the same line, and inconsistent equations form parallel lines with no point of intersection.

Example A.3   A manufacturer has 10 packages of part X and 12 packages of part Y in stock. The total cost for these units was $548. Additional supplies are purchased, 4 packages of part X and 3 packages of part Y, for a total cost of $170. Determine the cost for parts X and Y.

Solution

Given:
The total cost and number of units for the two parts X and Y. Purchases were made on two separate occasions.

Find:
The unit cost for X and Y.

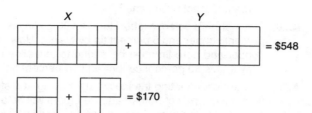

Assumption:

The unit price did not change.

Analysis:

In this case we know the total cost, but not the individual cost of X and Y. Let

$$x = \text{cost of 1 unit of } X$$

$$y = \text{cost of 1 unit of } Y$$

The total cost of the purchases was

$$10x + 12y = 548 \qquad \textbf{(A.1)}$$

$$4x + 3y = 170 \qquad \textbf{(A.2)}$$

To solve by addition, multiply Equation (A.2) by 4 and subtract this from Equation (A.1):

$$10x + 12y = 548$$

$$\underline{-16x - 12y = -680}$$

$$-6x = -132$$

$$x = 22$$

Substitute back in either Equation (A.1) or (A.2) and solve for y. For instance, substituting in Equation (A.1) yields

$$10(22) + 12y = 548$$

$$y = 27.33$$

A.26. A computer manufacturer ships 200 computers to two different stores, A and B. It costs $4.50 to ship to A and $3.75 to ship to B. The total shipping invoice was $806.25. How many computers were shipped to each location?

Problems

A.27. Tank *A* contains a mixture of 100 liters (L) of water and 50 L of alcohol while tank *B* has 120 L of water and 30 L of alcohol. How many liters should be taken from the tanks to create an 80-L mixture that is 25 percent alcohol by volume?

A.28. In a materials science laboratory, a 100-g alloy is found to contain 20 percent copper and 5 percent tin by weight. How many grams of pure copper and pure tin must be added to this alloy to produce another alloy that is 30 percent copper and 10 percent tin?

A.29. You are coordinating the ice cream making at a state agricultural fair where two types of homemade ice cream will be available. There are two main ingredients, eggs and cream, and 500 eggs and 900 cups of cream are available. Plain vanilla requires 1 egg and 3 cups of cream per quart while French vanilla needs 2 eggs and 3 cups of cream per quart. Determine the number of quarts of each variety that should be made to use up all the ingredients.

A.30. In intramural sports, one dorm has won a total of 12 games this year, some in volleyball and others in soccer. There is a rating system in which each win in volleyball counts as 2 points and each win in soccer counts as 4 points. The dorm has a total of 38 points. How many soccer and volleyball games did it win?

A.31. As a project manager, you are responsible for allocating a $10,000 bonus among four full-time and two part-time employees. You decide on the algorithm that the full-time employees will receive an amount which is twice that of the part-timers. What are the amounts the full-timers and part-timers receive?

Linear Systems with Three Variables

It is possible to extend the technique of solving two simultaneous equations to three or more simultaneous equations. Usually when three or more simultaneous equations are solved, computers assist in the process. However, on occasion it is necessary to perform the operations by hand, and you certainly should understand the manipulations that the software is performing.

Consider the following set of equations:

$$2x + y - z = 2 \qquad \textbf{(A.3)}$$

$$x + 3y + 2z = 1 \qquad \textbf{(A.4)}$$

$$x + y + z = 2 \qquad \textbf{(A.5)}$$

The values of x, y, and z that simultaneously satisfy the equations may be found by addition and substitution. The next section, on matrices, will discuss a more general method of solution. In this case, however, first add Equations (A.3) and (A.5), eliminating the variable z, yielding

$$3x + 2y = 4 \qquad \textbf{(A.6)}$$

Eliminate z from another set of two equations, say (A.3) and (A.4), by multiplying (A.3) by 2 and adding the equations, which yields

$$5x + 5y = 5$$

$$x + y = 1 \qquad \text{(A.7)}$$

Solve Equations (A.6) and (A.7) simultaneously, substituting x in terms of y from Equation (A.7) into Equation (A.6).

$$3(1 - y) + 2y = 4$$

$$y = -1$$

From Equation (A.7),

$$x = 1 - y = 1 - (-1) = 2$$

Solve for z by substituting into one of the original equations. For instance, picking Equation (4.5) yields

$$2 - 1 + z = 2$$

$$z = 1$$

Thus, the simultaneous solution of three independent equations is $x = 2$, $y = -1$, $z = 1$. An important habit to develop is to verify the results. Substitute the answers back into the original equation, making sure they satisfy the problem. There are many applications of simultaneous equations in engineering, so facility in both developing and solving them is very important.

Example A.4   An agricultural engineer is developing a new animal feed which should have 22 kg of protein, 28 kg of fat, and 18 kg of fiber. There are three types of plants available: corn, cottonseed, and soybeans. The table below indicates the kilograms of protein, fat, and fiber per kilogram of plant food. The percentages do not total to unity, as the plant contains more nutrients than are listed.

	Corn	Cottonseed	Soybeans	Total, kg
Protein, kg	0.25	0.2	0.4	22
Fat, kg	0.4	0.3	0.2	28
Fiber, kg	0.3	0.1	0.2	18

Solution
Given:
The requirements for an animal feed in terms of protein, fat, and fiber as well as percentages these components make up in the three available plant foods used to make the animal feed.

Find:
The kilograms of each plant food used to make the animal feed.

Sketch and Data:

Corn	Cottonseed	Soybean	Total
25% protein	20% protein	40% protein	22 kg
40% fat	30% fat	20% fat	28 kg
30% fiber	10% fiber	20% fiber	18 kg
x	y	z	

Figure A.4

Assumptions:
None.

Analysis:
Let x = kilograms of corn, y = kilograms of cottonseed, and z = kilograms of soybeans required. The 22 kg of protein must be made by combining

$$0.25x + 0.2y + 0.4z = 22 \qquad \textbf{(A.8)}$$

The 28 kg of fat is found by combining

$$0.4x + 0.3y + 0.2z = 28 \qquad \textbf{(A.9)}$$

The 18 kg of fiber is found by combining

$$0.3x + 0.1y + 0.2z = 18 \qquad \textbf{(A.10)}$$

Subtract Equation (A.10) from (A.9), yielding

$$0.1x + 0.2y = 10$$

$$x + 2y = 100 \qquad \textbf{(A.11)}$$

Multiply Equation (A.9) by 2, and subtract Equation (A.8) from it, which yields

$$0.55x + 0.4y = 34 \qquad \textbf{(A.12)}$$

Solve Equation (A.11) for x in terms of y, and substitute into Equation (A.12), which yields

$$0.55(100 - 2y) + 0.4y = 34$$

$$-0.7y = -21$$

$$y = 30 \text{ kg}$$

From Equation (A.11), $x = 40$ kg and from Equation (A.10),

$$0.3(40) + 0.1(30) + 0.2z = 18$$

$$z = 15 \text{ kg}$$

Thus the feed should contain 40 kg of corn, 30 kg of cottonseed, and 15 kg of soybeans.

Matrix Solution of Linear Systems

There are several ways in which matrices may be used to solve linear systems. What follows describes one method. Consider a set of three equations and three unknowns x, y, and z.

$$a_1x + b_1y + c_1z = d_1$$

$$a_2x + b_2y + c_2z = d_2$$

$$a_3x + b_3y + c_3z = d_3$$

This set of equations may be written in an abbreviated form as

$$\left. \begin{array}{ccc} a_1 & b_1 & c_1 \\ a_2 & b_2 & c_2 \\ a_3 & b_3 & c_3 \end{array} \right| \begin{array}{c} d_1 \\ d_2 \\ d_3 \end{array}$$

This rectangular array of numbers is called a *matrix*. Each number is called an *element* in the array and is identified by its row and column numbers. Thus, b_1 is in element $(1, 2)$. The last column is separated from the first three by a vertical line; this indicates that it augments the matrix formed by the equations' coefficients.

Consider the following set of equations:

$$\begin{array}{rcrcrcl} x & - & 2y & + & z & = & 5 \\ -2x & + & 4y & - & 2z & = & 2 \\ 2x & + & y & - & z & = & 2 \end{array}$$

The augmented matrix is

$$\left. \begin{array}{ccc} 1 & -2 & 1 \\ -2 & 4 & -2 \\ 2 & 1 & -1 \end{array} \right| \begin{array}{c} 5 \\ 2 \\ 2 \end{array}$$

For any augmented matrix of a set of linear equations, the following transformations result in an equivalent matrix:

1. Any two rows may be interchanged.
2. The elements of any row may be multiplied by a nonzero real number.
3. Any row may be changed by adding to its elements a multiple of the corresponding elements of another row.

The Gauss-Jordan method reduces the augmented system matrix to a unity matrix of the form

$$\begin{array}{ccc|c} 1 & 0 & 0 & a \\ 0 & 1 & 0 & b \\ 0 & 0 & 1 & c \end{array}$$

where a, b, and c become the solutions of x, y, and z that solve the set of simultaneous equations.

Consider first a linear system of two equations

$$3x - 4y = 1$$

$$5x + 2y = 19$$

The augmented matrix is

$$\begin{array}{cc|c} 3 & -4 & 1 \\ 5 & 2 & 19 \end{array}$$

Multiply each element in row 1 by 1/3 to obtain a one in element (1, 1).

$$\begin{array}{cc|c} 1 & -\dfrac{4}{3} & \dfrac{1}{3} \\ 5 & 2 & 19 \end{array}$$

To obtain a 0 in element (2, 1), multiply each element of row 1 by -5 and add it to the element in row 2.

$$\begin{array}{cc|c} 1 & -\dfrac{4}{3} & \dfrac{1}{3} \\ 0 & \dfrac{26}{3} & \dfrac{52}{3} \end{array}$$

To obtain a 1 in element (2, 1), multiply each element in row 2 by $\frac{3}{26}$.

$$\begin{array}{cc|c} 1 & -\dfrac{4}{3} & \dfrac{1}{3} \\ 0 & 1 & 2 \end{array}$$

To obtain a 0 in element (1, 2), multiply each element of row 2 by 4/3 and add this value to the corresponding element in row 1.

$$\begin{array}{cc|c} 1 & 0 & 3 \\ 0 & 1 & 2 \end{array}$$

Thus, the solution of the two original equations is $x = 3$, $y = 2$. This may be extended to a system with three equations. Obviously, it is time-consuming, but it is procedural and as long as one is careful, the solution to sets of equations may be determined. Many calculators as well as most spreadsheets have matrix equation solvers built in; the coefficients of the augmented matrix need only be entered.

A.32. Tastee Beverage Company makes three types of juice drinks: Cran-Orange, using 1 quart (qt) of cranberry juice and 3 qt of orange juice; CranPine, using 1 qt of cranberry juice and 3 qt of pineapple juice; and PineOrange, using 2 qt of pineapple and 2 qt of orange juice per gallon. Each day the company uses 350 qt of cranberry juice, 800 qt of orange juice, and 650 qt of pineapple juice. How many gallons of each blend must be produced daily to use the above amounts of juice?

A.33. As president of ExpressAir, you are considering the purchase of additional airplanes to expand your company's capacity by 2000 seats. A mix of aircraft type is desired because of routing, and the following information is known: Boeing 747s cost $150 million each and carry 400 passengers, Boeing 777s cost $115 million each and carry 300 passengers, and Airbus A321s cost $60 million and carry 200 passengers. The routes indicate that a wise mix of aircraft would be equal numbers of 747s and 777s. The total budget available is $710 million. How many of each aircraft can be purchased and still satisfy the seating increase?

A.34. A Florida company maintains two distribution warehouses, one in Jacksonville and the other in Sarasota. The warehouses supply software manuals to two retail outlets, one in Orlando and the other in Tallahassee. The Jacksonville warehouse has 1000 manuals, and the one in Sarasota has 2000 manuals. Each retail store orders 1500 manuals. It costs $1 to ship a manual from Jacksonville to Orlando and $2 to ship one from Sarasota to Orlando. It costs $5 to ship a manual from Jacksonville to Tallahassee and $4 to ship a manual from Sarasota to Tallahassee. For a budget of $9000, how many manuals should be shipped from each warehouse to satisfy each store's requirements?

A.35. Solve Example A.4, using matrix methods.

A.36. A service station sells three grades of gasoline: regular, premium, and super. One day the station sold 150 gal of regular, 400 gal of premium, and 130 gal of super for a total of $909. The next day it sold 170 gal of regular, 380 gal of premium, and 150 gal of super for $931. The price difference between super and regular is one-half the difference between premium and regular grades. Determine the cost per gallon for each grade of gasoline.

A.37. A chemical engineer has three salt solutions available, 5 percent, 15 percent, and 25 percent, to make 50 L of a 20 percent solution. There is much more 5 percent solution available, so a requirement is to use twice as much 5 percent solution as 15 percent solution. Determine the amount of each salt solution that is used to make the mixture.

A.38. A manufacturing company produces two products, I and II, that require time on machines A and B. Product I requires 1 h on A and 2 h on B, while product II requires 3 h on A and 1 h on B. The company is open 16 h/day, with the machines operating 15 h/day. What is the number of each product that can be produced daily?

A.39. Three grades of resin are available which may be mixed together to form a fourth resin. The costs of the initial resins are $4.60, $5.75, and $6.50 per pound. The mixture value will be $5.25 per pound, and 20 pounds (lb) is needed. In addition, the amount of the least expensive resin should be equal to the total amount of the other two. Determine the amount of each resin needed.

A.40. An engineering club holds a benefit party and collects a total of $2480, consisting of $5, $10, and $20 bills. The total number of bills is 290. The value of the total number of $10 bills is $60 more than the value of the total number of $20 bills. Determine the number of each type of bill the club has.

Quadratic Equations with One Unknown

A quadratic equation has the form $ax^2 + bx + c = 0$, where x is the variable and a, b, and c are constants with a not equal to zero. There are several methods used to solve this equation, such as factoring, completing the square, graphs, and the quadratic formula. The quadratic formula is

$$x = \frac{-b \pm \sqrt{b^2 - 4ac}}{2a}$$

There will be two roots to a quadratic equation; two values of x will satisfy the equation. Often only one is physically possible, and this is the selected value.

Example A.5   A supplier bought a certain number of copies of software for $1800 and sold all but six copies. The supplier made a profit of $20 per copy for each package of software sold and decided to reinvest the revenue in more copies. The resultant purchase was 30 copies more than the original order. Find the unit software cost.

Solution

Given:
The total cost for an unknown amount of software, the profit for the copies sold, and the number of additional copies that can be purchased.

Find:
The unit cost of the software.

Sketch and Data:

Total cost = $1800
x = cost/copy

Figure A.5

Assumptions:

The unit cost of the software did not change.

Analysis:

Let x be the cost per copy of software purchased for $1800. The total number of copies is $1800/x$. The total revenue TR received is

(Number of copies sold) $\times$ (Income per copy) $=$ Total revenue

$$\left(\frac{1800}{x} - 6\right)(x + 20) = \text{TR} \qquad \textbf{(A.13)}$$

This value of total revenue may also be represented by the total number of new copies of software purchased.

(Number of copies purchased) $\times$ (Cost per copy) $=$ Total revenue

$$\left(\frac{1800}{x} + 30\right)(x) = \text{TR} \qquad \textbf{(A.14)}$$

Equating Equations (A.13) and (A.14) yields

$$\left(\frac{1800}{x} - 6\right)(x + 20) = (x)\left(\frac{1800}{x} + 30\right)$$

$$36x^2 + 120x - 36{,}000 = 0$$

$$x^2 + 3.333x - 1000 = 0$$

Use the quadratic formula to solve for x.

$$x = \frac{-3.333 \pm \sqrt{11.111 - (4)(1)(-1000)}}{2(1)}$$

$$= 30, -33.33$$

The only physically possible answer is $30 per copy.

A.41. A right triangle is formed from a wire 60 cm long. The triangle's hypotenuse is 25 cm. Find the lengths of the other two sides.

A.42. A student is given a 9-in. by 12-in. piece of paper and is told to construct an open box by cutting equal squares from each of the corners of the paper and then folding up the sides. The base areas should be 60 in^2. Find the length of the sides of the squares that are removed.

A.43. A student is driving home, a distance of 150 mi, for the weekend. From previous experience, the student knows that increasing the average speed by 10 mi/h could reduce the time of the trip by 35 min. What is the actual average speed?

A.44. Two pipes, I and II, can be used to fill a tank. Pipe I fills the tank in 4 h. If pipe II is used by itself, it takes 3 h longer than if both pipes are

Problems

used simultaneously. Determine the time it takes to fill the tank with pipe II.

Exponential and Logarithmic Functions

An exponential function may be defined as $f(x) = a^x$. Examples of this are $a^{2.5}$, 2^x, and 3^{-1}. Exponential growth is frequently encountered in engineering, science, and mathematics. Imagine that a population is increasing at 6 percent per year. If the initial population is I_0, then the population after n years is $I = I_0(1 + 0.06)^n$. If a certain animal species increases at a rate of 6 percent and its initial population is 10,000, the population after 5 years is $I = (10,000)(1.06)^5 = 13,382$.

A very useful exponential function occurs when $f(x) = e^x$, in which e is the irrational number 2.718281828. The exponential function is used to model many natural systems, such as the exponential decay of radiation, transient responses in electrical and mechanical systems, and population change.

A logarithm is the exponent of an exponential function that represents a real, positive number. Two logarithms are commonly used: logs to the base 10 (common logarithms) and logs to the base e (natural logarithms). Consider the number 820 represented as 10^x, where x is 2.9138, or e^x where x is 6.7093. The exponent is the logarithm of the number to a given base. Thus,

$$\log_{10} 820 = 2.9138$$

$$\log_e 820 = \ln 820 = 6.7093$$

An example of a log function commonly used is the decibel rating of sound. The loudness of sound is measured in decibels (dB) and is a $\log_{10}$ equation (we omit the subscript 10, as it is understood):

$$dB = 10 \log(I/I_0)$$

where I_0 is the intensity of the faintest sound that can be heard and I is the intensity of the actual sound. If $I = 10,000 I_{10}$, then

$$dB = 10 \log_{10}(10,000 I_0/I_0) = 40$$

For the decibels to increase from 40 to 50, the intensity must increase from $10,000 I_0$ to $100,000 I_0$.

Sometimes students have difficulty solving exponential and logarithmic equations. Consider the equation

$$50 = 1000(1 - e^{-3t})$$

and solve it for t. The following steps may be used:

$$0.05 = 1 - e^{-3t}$$

$$0.95 = e^{-3t}$$

$$\ln 0.95 = -0.05129 = -3t \ln e = -3t(1)$$

$$t = 0.017$$

Solve the following equation for x:

$$y = a \ln(1 + x/a)$$

$$y/a = \ln(1 + x/a)$$

$$e^{y/a} = 1 + x/a$$

$$x = a(e^{y/s} - 1)$$

Problems

A.45. Assume the annual rate of inflation is 5 percent. Determine how long it will take for prices to double if they rise in proportion to inflation.

A.46. The half-life of radioactive carbon 14 is 5700 years. After a plant or animal dies, the level of carbon 14 decreases as the radioactive carbon disintegrates. The decay of radioactive material is given by the relationship $A = A_0 e^{-kt}$, where A_0 is the initial amount of material at time 0 and t represents the time measured from time 0 in years. For carbon 14, $k = 1.216 \times 10^{-4}$ years. Samples from an Egyptian mummy show that the carbon 14 level is one-third that found in the atmosphere. Determine the approximate age of the mummy.

A.47. Paint from cave drawings in France indicates a carbon 14 level 15 percent of that found in the atmosphere. Determine the approximate age of the drawings.

A.48. The amount of a certain chemical A that will dissolve in solution varies exponentially with the Celsius temperature T according to the equation $A = 10e^{0.01T}$. Determine the temperature that allows 15 g of chemical to dissolve.

A.49. Newton's law of cooling describes the cooling or heating of an object by a fluid (liquid or gas). The temperature variation with time is given by the equation $T(t) = T_0 + Ae^{-kt}$, where A is a constant equal to 100, k is a constant equal to 0.1, t is the time in minutes, and T_0 is the surrounding fluid temperature. Determine the time it will take a cup of hot coffee to cool to 30° in a room at 20°.

A.50. Plutonium 239 decays at a rate of 0.00284 percent per year. If the initial sample size of P-239 is 10 g, how much will remain as P-239 after 20,000 years?

A.51. A strain of bacteria is reproducing continuously at a rate of 0.31 percent per minute. A culture with 1000 organisms will double in size in what amount of time?

A.52. The Richter scale is used to measure the intensity of earthquakes and is given by the formula $R = 0.667(\log E - 4.4)$, where E is the energy released in an earthquake, measured in joules. The San Francisco earthquake of 1906 registered 8.2 on the Richter scale, and one in 1989 measured 7.1. What is the percentage of energy released in the 1989 earthquake compared to the one in 1906?

A.53. The decibel level of sound from a stereo set decreases with distance according to the relationship

$$Db = 10 \log \left(\frac{320 \times 10^7}{r^2} \right)$$

Determine the decibel rating at 5, 10, and 15 ft. Express the relationship in the form $Db = a + b \log r$.

Trigonometry

Algebra and trigonometry are related in that trigonometry problems often require algebraic problem-solving ability. Figure A.6 illustrates a right triangle. The Pythagorean theorem is extensively used in the solution of right triangles: $a^2 + b^2 = c^2$. The commonly used trigonometric functions are as follows:

Trigonometric ratio	Abbreviation	Definition	
sine of θ	$\sin \theta$	$\dfrac{\text{Side opposite}}{\text{Hypotenuse}}$	$= \dfrac{a}{c}$
cosine of θ	$\cos \theta$	$\dfrac{\text{Side adjacent}}{\text{Hypotenuse}}$	$= \dfrac{b}{c}$
tangent of θ	$\tan \theta$	$\dfrac{\text{Side opposite}}{\text{Side adjacent}}$	$= \dfrac{a}{b}$

Angles are measured in degrees or radians. There are 360° in a circle and 2π radians (rad). Figure A.7 illustrates angles

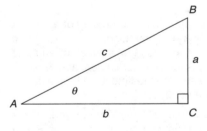

Figure A.6

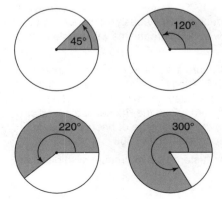

Figure A.7

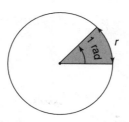
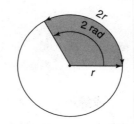

using degree notation, and Figure A.8, degree measurements with radian notation. Note that 1 rad is the angle whose subtended arc has a length equal to the radius of the circle. Each degree is equal to $\pi/180$ rad. Calculators use either degrees or radians; be sure to check which is required. Computer programs often use the trigonometric angles measured in radians.

Problems

A.54. A plot of land is a 270° sector with a 10-ft radius. Determine the area.

A.55. A curve along a highway is a circular arc 50 m long with a radius of curvature of 250 m. How many degrees does the highway change its direction along the arc?

A.56. Find the area of the sector inside the square *ABCD*.

A.57. The angle of elevation to the top of a flagpole is 40° from a point 30 m from the base of the pole. What is the height of the pole?

A.58. A kite string forms an angle of 42° with the ground when the entire 800 ft of string is used. What is the kite's elevation?

A.59. An 80-ft pole is stabilized by guide wires which run from the top of the pole to the ground. The wires are located 15 ft from the base of the pole. What length of wire is required? What is the angle that the wire makes with the ground?

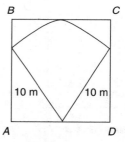

Figure AP.56

In general, many triangles are not right triangles, but obtuse triangles, as illustrated in Figure A.9. There are two laws that assist in providing relationships between sides and angles. The *law of cosines* is

Laws of Cosines and Sines

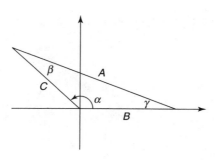

Figure A.9

$$A^2 = B^2 + C^2 - 2BC \cos \alpha$$

and the *law of sines* is

$$\frac{A}{\sin \alpha} = \frac{B}{\sin \beta} = \frac{C}{\sin \gamma}$$

Problems

A.60. A surveyor measures the angle of elevation of a mountain from point A and finds it to be 23°. The surveyor moves 1/4 mi closer to the mountain and finds the angle of elevation is 43°. What is the height of the mountain?

A.61. A 12-ft flagpole stands at the edge of a building's roof. The angle of elevation from the ground 65 ft from the building to the top of the flagpole is 78°. Determine the building's height.

A.62. A diagonal of a parallelogram has length of 60 in. and makes an angle of 20° with one of the sides. The side has a length of 25 in. Determine the length of the other side of the parallelogram.

A.63. Determine the length of AB in Figure AP.63.

A.64. A surveyor is determining the distance between two points A and B located on the shoreline. The surveyor is located at point C and measures the distance AC to be 180 m and BC to be 120 m. The angle at C is 56°. Find the distance AB.

A.65. In Problem A.64, let the angle at C be 130°, and find AB.

A.66. Two guide wires are attached to the top of a pole and are anchored into the ground on opposite sides of the pole at points A and B. The ground is the same elevation relative to the pole in all directions. The distance AB is 40 m, and the angles of elevation at A and B are 70° and 55°, respectively. Determine the guide-wire lengths.

A.67. An airplane is flying in a straight line and at constant elevation toward an airfield. At a given instant, the angle of depression between the plane and airfield is 32°. After it flies two miles, the angle of depression is 74°. What is the distance between the plane and the airfield at the second point?

A.68. In Figure AP.68, points A and B are on the same side of the river, and the distance AB is 600 ft. Determine the distance CD on the opposite side of the river.

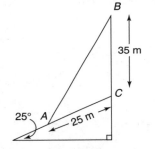

Figure AP.63

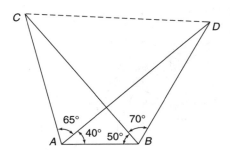

Figure AP.68

All number systems have a certain commonality of structure, and before we analyze the binary system, we first examine the decimal, or base 10, system.

Consider the number 542.1. It is really a group of additions of numbers of various powers of 10:

$$
\begin{array}{ll}
500 & 5 \times 10^2 \\
40 & 4 \times 10^1 \\
2 & 2 \times 10^0 \\
\underline{0.1} & 1 \times 10^{-1} \\
542.1 &
\end{array}
$$

The value of the right-hand column can be generalized with the coefficients of the powers written as A_n; thus 5 is the coefficient of 10^2, and any number in decimal would have the decimal form of

$$A_n \times 10^n + A_{n-1} \times 10^{n-1} + \cdots + A_2 \times 10^2 + A_1 \times 10^1 + A_0 \times 10^0$$

We can use the same concepts inherent in the previous equation for any number system. Let the number system be the binary number system, where numbers are to the base 2. Thus, any number can be written as

$$B_n \times 2^n + B_{n-1} \times 2^{n-1} + \cdots + B_2 \times 2^2 + B_1 \times 2^1 + B_0 \times 2^0$$

In the decimal system A_n can assume any value between 0 and 9, but 10 is not an allowed coefficient. If we extend this to the binary system, B_n can assume any value between 0 and 1, and 2 is not an allowed coefficient.

Representations and Calculations

The following convention is used to distinguish between number systems. Let us write the numbers 2 and 10 in each of their number systems. In the decimal system 2 is 2×10^0, whereas in the binary system 2 is $1 \times 2^1 + 0 \times 2^0 = 10_2$, where the subscript 2 tells us that this the base-2 system.

The decimal system expresses the number 10 as

$$1 \times 10^1 + 0 \times 10^0 = 10$$

In the binary system 10 is expressed as

$$
\begin{array}{ccccccccc}
1 \times 2^3 & + & 0 \times 2^2 & + & 1 \times 2 & + & 0 \times 2^0 & = & 1010_2 \\
8 & + & 0 & + & 2 & + & 0 & = & 10_{10}
\end{array}
$$

Consider the opposite situation: convert 1001110_2 to its decimal equivalent. Count the number of places in the expression to determine how many terms are involved in the conversion. In this case there are seven.

$$1 \times 2^6 + 0 \times 2^5 + 0 \times 2^4 + 1 \times 2^3 + 1 \times 2^2 + 1 \times 2^1 + 0 \times 2^0$$
$$64 \quad + \quad 0 \quad + \quad 0 \quad + \quad 8 \quad + \quad 4 \quad + \quad 2 \quad + 0 = 78_{10}$$

Octal and Hexadecimal Systems

Two other number systems are often used in computers: the octal, or base-8, system, and the hexadecimal, or base-16, system. In the octal system, the range of the coefficients is from 0 to 7. In the hexadecimal system a problem arises if we let the coefficients vary from 0 to 15, as any value beyond 9 is not unique. To resolve this, the hexadecimal system uses 0 to 9 plus A to F to uniquely define the coefficients. Table A.1 gives the equivalents for the number systems. Conversion between decimal and octal or hexadecimal is rare. Much more frequent, and simpler, is the conversion between the binary system and the octal or hexadecimal system. In this case, the conversion is much simpler, as 8 and 16 are powers of 2.

Use Table A.1 to convert 10100110_2 to hexadecimal notation. Divide the term into groups of four, starting with the right-hand side and adding zeros as necessary to the left-hand side to create the necessary groupings. Thus, the three groups are

$$0001_2 \quad 0011_2 \quad 0101_2$$
$$1 \qquad 3 \qquad 5 \qquad = 135_{16}$$

Table A.1 Decimal, binary, octal, and hexadecimal number equivalents

Decimal	Binary	Octal	Hexadecimal
0	0	0	0
1	1	1	1
2	10	2	2
3	11	3	3
4	100	4	4
5	101	5	5
6	110	6	6
7	111	7	7
8	1000	10	8
9	1001	11	9
10	1010	12	A
11	1011	13	B
12	1100	14	C
13	1101	15	D
14	1110	16	E
15	1111	17	F
16	10000	20	10
17	10001	21	11
18	10010	22	12
19	10011	23	13
20	10100	24	14

Operations that are easy in the decimal system become more complicated in the binary system. In part this is because our mind-set is conditioned to the decimal world.

Addition

Consider the following addition operations in binary: $0 + 0 = 0; 0 + 1 = 1; 1 + 1 = 0 = 10_2$. In this example, 1 is carried to the next place, leaving a 0 in the original location. To help with additions, we can use a truth table, as shown in Table A.2. It helps to remember terms from arithmetic; *augend* is a number to which another number is added, while *addend* is a number that is added to another. The result is a *sum* and/or a *carry* to the next place. The table represents all the possible combinations that can occur. Caution is required when you carry a 1 in addition, as it may cause an additional 1 to carry.

For instance, add the binary numbers 1 and 1010 as follows:

$$\begin{array}{r} 1010 \\ \underline{0001} \\ 1011 \end{array}$$

If binary numbers 1 and 1111 are added, we get the following results:

$$\begin{array}{r} 1111 \\ \underline{0001} \\ 10000 \end{array}$$

The carry created 0s, and an additional place was required.

Table A.2   Binary addition truth table

Augend	Addend	Sum	Carry
0	0	0	0
0	1	1	0
1	0	1	0
1	1	0	1

Subtraction

Subtracting binary numbers produces another truth table, Table A.3. The *subtrahend* is the number to be subtracted, the *minuend* is the number from which the subtrahend is subtracted, yielding the *difference* between them. At times it is necessary to *borrow* from the next place. Subtract the following binary numbers. The italic *1* over the minuend indicates that a borrow is necessary.

$$
\begin{array}{cc}
1 & 1111 \\
11101 & 10000 \\
-\,1011 & -\quad 11 \\
\hline
10010 & 1101
\end{array}
$$

Actually, computers subtract by creating a complement of the subtrahend and adding this to the minuend. Thus, both addition and subtraction become processes of addition. We will see that multiplication and, by extension, division also become processes in addition that the computer quickly performs .

Table A.3 Binary subtraction truth table

Minuend	Subtrahend	Difference	Borrow
0	0	0	0
0	1	1	1
1	0	1	0
1	1	0	0

Multiplication

Multiplication is an addition process; if we are multiplying 18×12, for instance, 18 is added to 0 twelve times or to itself 11 times. In the process of multiplication in the decimal system, we first multiply 18×2 and add this to the multiplication of 18×10. In the binary system a similar process occurs. First, the binary number system multiplication table is as follows:

$$0 \times 0 = 0$$

$$0 \times 1 = 0$$

$$1 \times 0 = 0$$

$$1 \times 1 = 1$$

Let us convert 18×12 to its binary equivalent of 10010×1100:

$$
\begin{array}{r}
10010 \\
\times\,1100 \\
\hline
00000 \\
00000 \\
10010 \\
10010 \\
\hline
11011000 = 216_{10}
\end{array}
$$

In a similar manner, division may be viewed as repeated subtractions, additions of the complement as noted, and this is the process for division as well.

A.69. Write the following decimal numbers in power-of-10 notation.
(*a*) 1990   (*b*) 22.22   (*c*) 0.1911

A.70. Write the following binary numbers in power-of-2 notation.
(*a*) 11001   (*b*) 100100111   (*c*) 0.11101

A.71. Convert the following decimal numbers to their binary equivalents.
(*a*) 296   (*b*) 16 941   (*c*) 3.1416   (*d*) 0.1110

A.72. Convert the following binary numbers to their decimal equivalents.
(*a*) 101.101   (*b*) 10001001   (*c*) 0.1110

A.73. Convert the following decimal numbers to their octal and hexadecimal equivalents.
(*a*) 19   (*b*) 2015   (*c*) 92.20   (*d*) 0.824

A.74. Convert the following binary numbers to their hexadecimal equivalents.
(*a*) 1010101101   (*b*) 1011.01   (*c*) 0.101101

A.75. Add the following pairs of binary numbers.

(*a*)
```
 1001
  101
```
(*b*)
```
 101
 111
```
(*c*)
```
 1101
   10
```
(*d*)
```
 10010010
 10001110
```

A.76. Subtract the following pairs of binary numbers.

(*a*)
```
 111
  11
```
(*b*)
```
 1001
  111
```
(*c*)
```
 1111
 1001
```
(*d*)
```
 1100011
 1011100
```

A.77. Multiply the following pairs of binary numbers.

(*a*)
```
 1101
  100
```
(*b*)
```
 10000
 10110
```
(*c*)
```
 110
  11
```
(*d*)
```
 110011
 100101
```

Answers to Selected Problems

Chapter 4

4.1. 23.5 Ω

4.3. 8.57 Ω

4.5. 13.57 Ω, 0.505 A, 0.379 A

4.7. 0.12 A

4.9. 1.25 Ω, 20 Ω

4.11. 2.96 Ω

4.13. $i_1 = -14$ A (from node); $i_2 = 16$ A; $i_3 = 6$ A; $i_4 = 22$ A

4.15. 20 V, 2 A

4.19. 9.807 V

4.21. 16 bulbs

4.23. 0.208 A

4.25. 8.02 Ω

4.27. 151.8 Ω

4.29. *a* does not balance; *b* essentially balances

4.31. Switch on AND bank door open AND safe door open AND alarm sounds

4.33.

A	B	C	D
0	0	0	0
0	1	0	1
1	0	0	1
1	1	1	0

4.37. Force *AC* = 13 252 N; Force *BC* = 9766 N

4.39. Drag = 53.7 N; *AC* = 128 N

4.41. *AC* = 321.3 N; *BC* = 212.2 N

4.43. *ACB* = 93.9 cm

4.45. $\Sigma M_A = 0$; $\Sigma M_B = 0$

4.47. $\Sigma M_A = 7$ N m

4.49. 1458.3 N; 1041.7 N

4.51. 159.4 lbf; 95.6 lbf

4.53. 7078 N

4.55. $\Delta L = 2.5$ mm; $\sigma = 177.8$ MPa

4.57. 222.8 MPa

4.59. 6.9 mm

4.61. 5.0 mm

4.63. 500 kg/s; 254.6 m/s

4.65. air 104.5°C; water 17.9°C

4.67. 22.4 C

4.69. (*a*) 2.79×10^7 kJ; (*b*) 37 453 kJ; (*c*) 287 250 kJ

4.71. (*a*) KE = 0, PE = 367.5 J; (*b*) KE = 245 J, PE = 122.5 J; (*c*) KE = 367.5 J, PE = 0

4.73. 187.5 kJ

4.75. $\Delta U = 68.6$ kJ

4.77. $\delta u = -481.6$ kJ/kg

4.79. 431.6 kJ

4.81. heat added = 30 000 kJ; heat rejected = 18 000 kJ

4.83. 10.24 m³/s

4.85. 4424 metric tons of coal daily; 5309 kg of sulfur daily; 360 railroad cars weekly

Appendix

- **A.1.** 100
- **A.3.** 150 gal
- **A.5.** 300 kg
- **A.7.** 55.5%
- **A.9.** 20 min
- **A.11.** 62.5 mph
- **A.13.** 16.92 in.
- **A.15.** 8.59
- **A.17.** 4 more workers
- **A.19.** 0.595 gal
- **A.21.** 62.8 mph
- **A.23.** 560 kN/m^2
- **A.25.** 16.4
- **A.27.** 30.77 lit, 49.23 lit
- **A.29.** 100 qts, 200 qts
- **A.31.** $1000
- **A.33.** 2-747's, 2-777's, 3-A321's
- **A.37.** 5 lit 15 percent, 10 lit 5 percent, 35 lit 25 percent
- **A.39.** 2 lbs, 8 lbs, 10 lbs
- **A.41.** 20, 15
- **A.43.** 46 mph
- **A.45.** 14.2 years
- **A.47.** 15,602 years
- **A.49.** 23 min
- **A.51.** 223.6 min
- **A.53.** $Db = 95.05 - 20 \log r$
- **A.55.** 11.4
- **A.57.** 25.17 m
- **A.59.** 162.8 ft
- **A.61.** 300.6 ft
- **A.63.** 48
- **A.65.** 273 ft
- **A.67.** 2.28 mi
- **A.71.** *(a)* 100101000
  - *(b)* 100001000101101
  - *(c)* 110010010001
  - *(d)* 0.00011100011
- **A.73.** *(a)* 23_8, 13_{16}
  - *(b)* 3737_8, $7DF_{16}$
  - *(c)* 134.14_8, $5C.3_{16}$
  - *(d)* 0.645_8, $0.D28_{16}$
- **A.75.** *(a)* 1110
  - *(b)* 1100
  - *(c)* 1111
  - *(d)* 100100000

Index